Roadside Geology of LOUISIANA

Darwin Spearing

1995
Mountain Press Publishing Company
Missoula, Montana

Library of Congress Cataloging-in-Publication Data

Spearing, Darwin.
Roadside geology of Louisiana / Darwin Spearing.
p. cm. — (Roadside geology series)
Includes bibliographical references (p. -) and index.
ISBN 0-87842-324-9 (pbk. : alk. paper)
1. Geology—Louisiana—Guidebooks. 2. Louisiana—Guidebooks.
I. Title. II. Series.
QE117.S64 1995 95-42705
557.63—dc20 CIP

Printed in the U.S.A.

Mountain Press Publishing Company
P. O. Box 2399 • Missoula, MT 59806
406-728-1900 • 800-234-5308

To my sons,
Dane and Craig,
scientist and artist,
respectively.

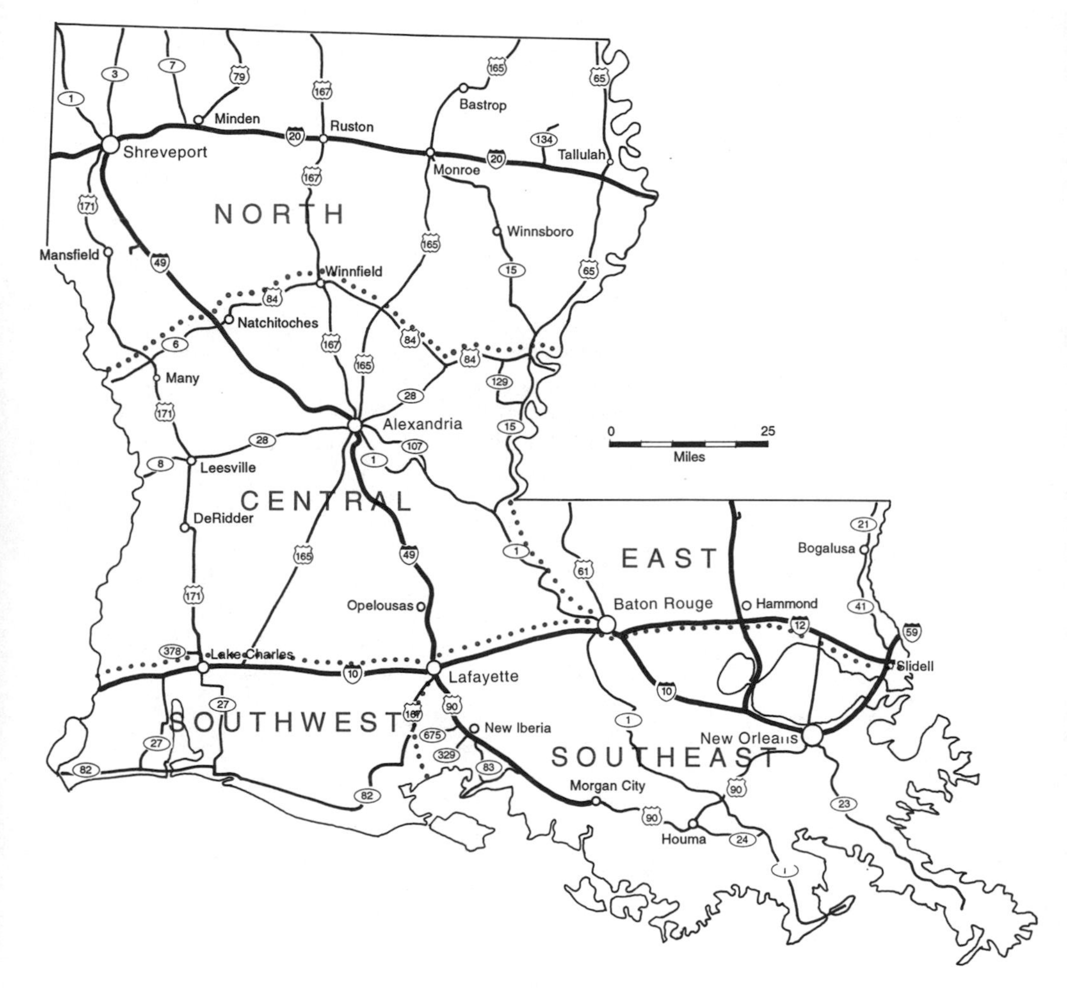

Roads and sections of Roadside Geology of Louisiana.

Contents

Preface

This book divides Louisiana into five regions, each a coherent package of land and rocks, each the subject of a chapter. This partitioning helps you find the roads you are looking for in the area you're in.

Each section begins with a map that shows the roads and rocks, followed by an introduction that sets the geological stage for the road descriptions and guides that follow.

An introductory section preceding the five regional chapters reviews the geologic history of Louisiana, and covers general topics such as salt domes, oil and gas, minerals, deltas, and rivers.

This book describes about 2,600 miles of paved roads. It covers all the interstate highways, as well as selected main highways that connect major towns or points of interest, such as state parks and significant geological or popular recreation areas. The purpose is to explain the geology of Louisiana as seen from your car window. This book is not a guide to the best geologic sites, which have a way of being remote riverbanks, old railroad cuts, inaccessible quarries, or along terrible dirt roads. Guidebooks are available to many of these special sites, and sources for them are listed in the back of the book.

If you travel roads not specifically covered, compare your state highway map to the geologic map at the head of the section in which you are driving. This will help you interpret much of what you see on the unguided road.

This book is written for people who are not geologists, but want to know something about the geology of Louisiana. Therefore, geologic words and professional jargon are minimized, but it is impossible to present geology without using at least some geologic terminology. If a term is unfamiliar, look it up in the glossary at the end of the book.

To cover the entire state requires omitting many details. This is not a guide to every single rock between points A and B. Main features and geologic highlights are emphasized to illustrate the principal geologic ideas exemplified along a particular road. Detailed road guides are available for many Louisiana roads. The section in the back of the book, "Maps and Information," will direct you to them.

One of the things you will need to develop is a keen sense of subtle changes in the landscape. The geology of Louisiana does not draw attention to itself the way rocks in such places as Colorado or New Mexico do. Plants cover most of the landscape, and the heavy rainfall effectively rounds the terrain and obliterates roadcuts and all but the most active stream banks. The rocks are still out there, though, expressed in the change in slope of a hill, a dip in the road, a shift in soil color, or an odd change of direction of a stream. Be alert to this gentle landscape; it tells its story, but you have to look carefully.

Geologic time is a consistent theme throughout the book. Geologic maps show areas where Miocene rocks, for example, are at the surface. Why such attention to the ages of rocks? The geologic history of the earth, like a giant jigsaw puzzle, must be assembled to be understood. If you know what happened in Miocene time in one place, then see different Miocene deposits in another place, a scenic panorama for that period of time develops in your mind. If you know the age, you can make comparisons from place to place that create a picture of the environment. This unfolding of ancient environments is one of the most exciting contributions geology has to offer the human mind. It is the only true experience we will ever have of traveling through time, science fiction notwithstanding!

The second major theme throughout is how the earth works. How do deltas form? Why do rivers meander? Louisiana is an excellent place to watch geologic processes that go on around us every day, processes that shape and mold the landscape, processes as dynamic in Louisiana as anywhere.

How can geologists examine an ancient sandstone in northern Louisiana and say, "This sand was deposited in a small, shallow stream"? They can because they have carefully studied many modern small, shallow streams and recognize the characteristics that distinguish their deposits. Louisiana has been the focus of intensive geological study of modern rivers, streams, barrier islands, deltas, swamps, natural river levees, and cheniers.

Geologists understand how the processes work in these environments and recognize the special features of the sediments that characterize each environment, whether the sediment was deposited yesterday or 40 million years ago. The fundamental idea is that geological processes, such as wind, flowing water, waves, and floods, worked in the past much like they do today. Being able to look at a flat river floodplain, a raised salt dome, or a watery marsh and envision the dynamic forces and changes that have shaped these landscapes is a distinct achievement of human understanding, one that I hope you will enjoy developing as you study the roadside geology of Louisiana.

Bon Voyage et Laissez les Bon Temps Rouler!

Acknowledgments

I thank a number of people who contributed to this book. Whitney Autin, Richard McCulloh, and John Snead, Louisiana State University, and David Alt and Donald Hyndman, University of Montana, kindly reviewed the draft manuscript and illustrations and made many helpful geological and editorial suggestions. Paul Albertson, Waterways Experiment Station, Corps of Engineers, Vicksburg, Mississippi, lent the book *Of Men and Rivers*, by Gary B. Mills, an intriguing historical account of the U.S. Army Corps of Engineers, Vicksburg District. Ervin Otvos generously sent several research papers.

Maps of the Old River control structures were kindly supplied by the people at Louisiana Hydroelectric, Vidalia, Louisiana. David Williamson, Central Louisiana Electric Company, Mansfield, Louisiana, cordially sent information on the Dolet Hills lignite mine. Thanks go to Charles Pearson, Whitney Autin, and the participants on the Friends of the Pleistocene 11th Annual Field Trip, for a wonderful, enlightening excursion. Thanks to Jack Cutshall, former state range conservationist, for his expertise on marshes and peat subsidence. *Merci* to Charles and Joan Escousse, New Orleans and Grand Lake, for the Louisiana culture lessons. Finally, to the fine folks at Mountain Press, hearty thanks for the cheerful support and hard work.

GEOLOGIC TIME SCALE

ERA	PERIOD	EPOCH	AGE (MILLIONS OF YEARS)
			0
CENOZOIC	QUATERNARY	Holocene (or "Recent")	0.1
		Pleistocene	2
	TERTIARY	Pliocene	5
		Miocene	24
		Oligocene	37
		Eocene	58
		Paleocene	66
MESOZOIC	CRETACEOUS		144
	JURASSIC		208
	TRIASSIC		245
PALEOZOIC	PERMIAN		286
	PENNSYLVANIAN		320
	MISSISSIPPIAN		360
	DEVONIAN		408
	SILURIAN		438
	ORDOVICIAN		505
	CAMBRIAN		570
PRECAMBRIAN	PROTEROZOIC		2500
	ARCHEAN		4500

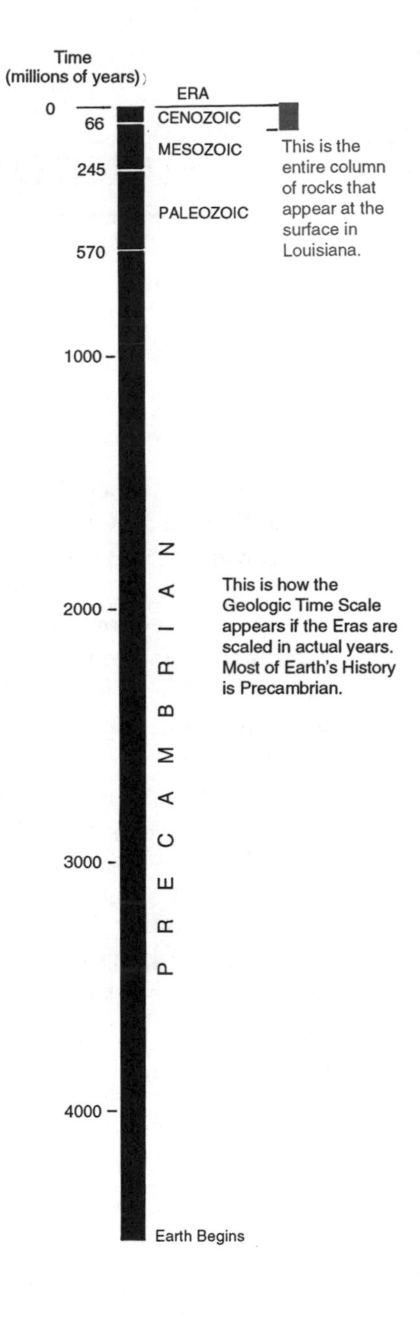

Geologic Overview

Geologic Time

The earth is old, very old indeed—more than 4 billion years old by current geologic reckoning, a reckoning, by the way, that has existed in human consciousness for less than 100 years. Before 1900, geologists had neatly divided up the earth's history into periods, the geologic time scale, based on major events they could read in the rocks and on the vertical sequence of fossils. But no one had discovered a way to measure the actual ages of rocks.

Some geologists tried to estimate the age of the oceans by estimating how long it would take for the world's rivers to deliver their inventory of salt, assuming the oceans started as fresh water. That did not work, and neither did a number of other early attempts to measure the age of the earth or its rocks. Despite the uncertainty of their estimates, geologists agreed before 1900 that the age of the earth was probably in the range of a few million to a few hundred million years.

The discovery of radioactivity and the development of radioactive age-dating techniques brought a new understanding of the age of the earth. The basic principles of radioactive age dating were understood shortly after the turn of the century, but the equipment necessary to date rocks did not become available until about fifty years later, when a flood of age dates left no doubt that the earth is several billion years old.

The geologic time scale divides earth history into packages of time that are easy to talk about. The Louann salt, now deep beneath Louisiana, was laid down in the early Gulf of Mexico during the Jurassic period, which was from 208 to 144 million years ago. If we want to consider another deposit or refer to events of the same age, we can simply describe them as Jurassic age. So the geologic time scale is a convenient shorthand to the geologic past.

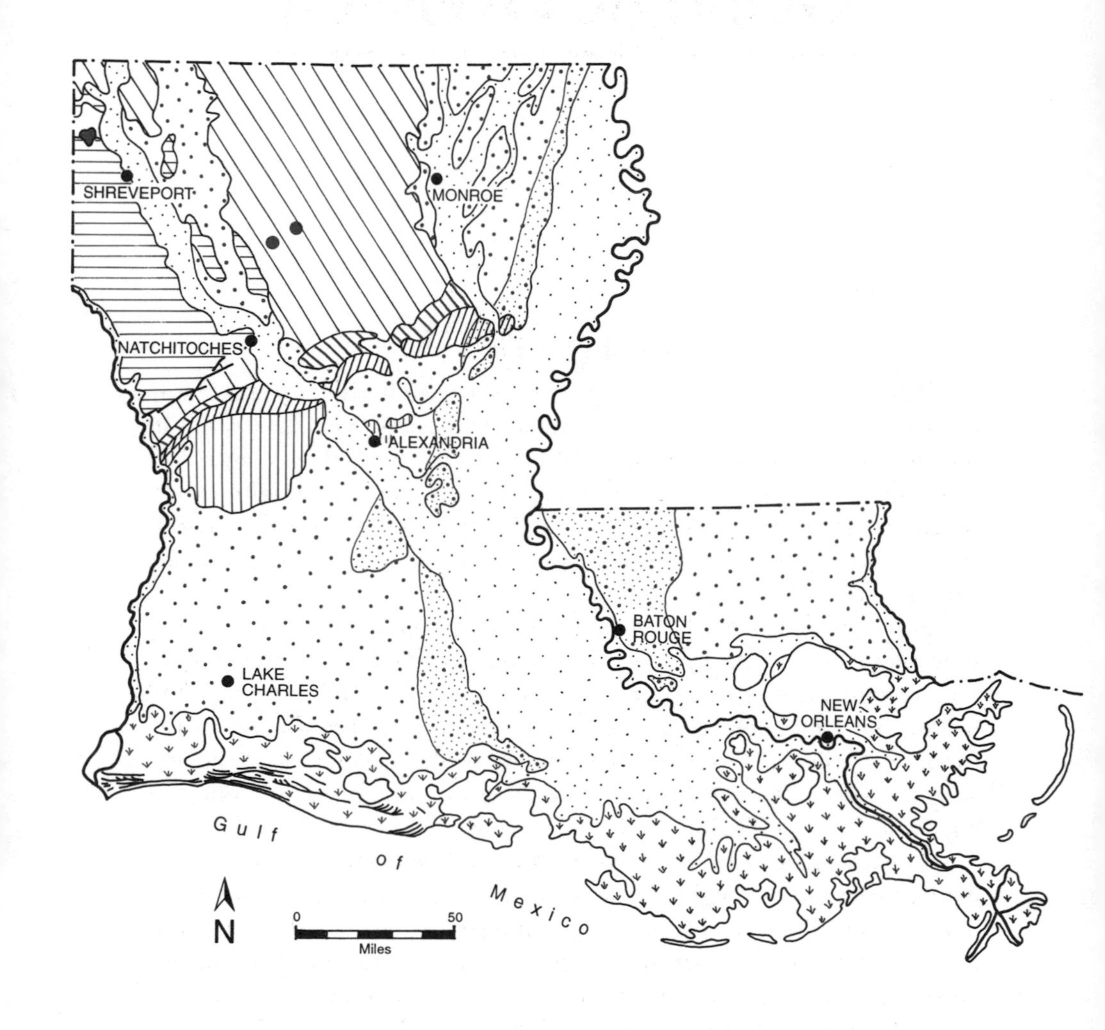

Geologic map of Louisiana.

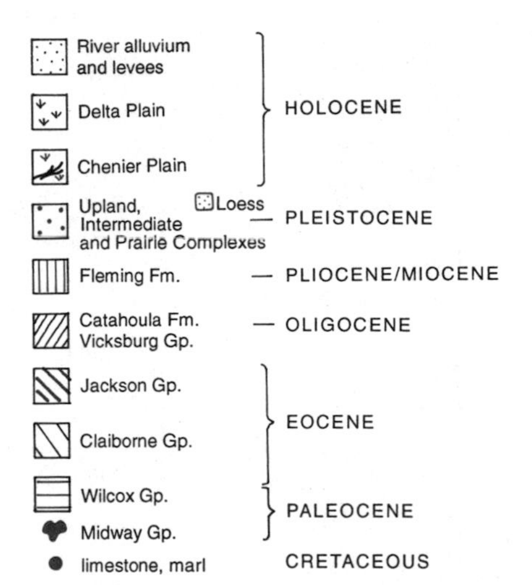
River alluvium and levees
Delta Plain
Chenier Plain
HOLOCENE
Upland, Intermediate and Prairie Complexes
Loess
PLEISTOCENE
Fleming Fm.
PLIOCENE/MIOCENE
Catahoula Fm.
Vicksburg Gp.
OLIGOCENE
Jackson Gp.
Claiborne Gp.
EOCENE
Wilcox Gp.
Midway Gp.
PALEOCENE
limestone, marl
CRETACEOUS

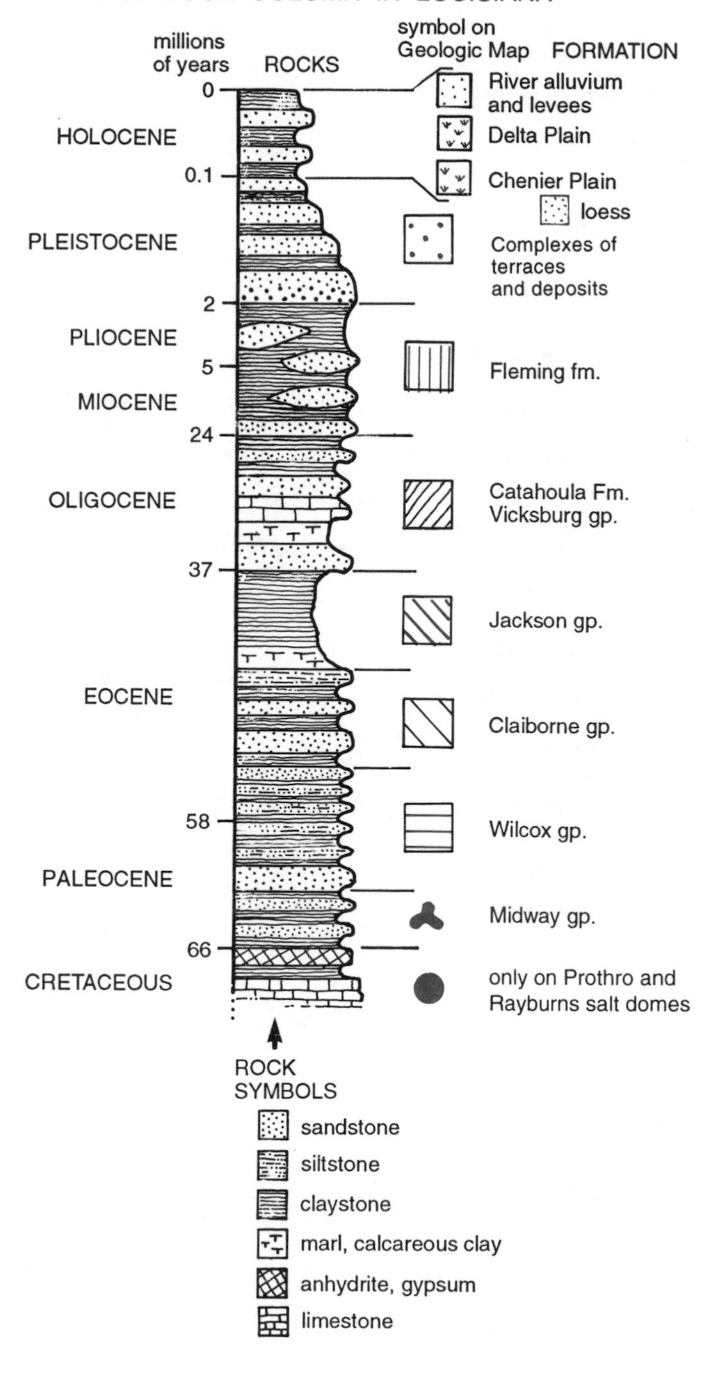
THE ROCK COLUMN IN LOUISIANA
millions of years
ROCKS
symbol on Geologic Map
FORMATION
0
0.1
2
5
24
37
58
66
HOLOCENE
PLEISTOCENE
PLIOCENE
MIOCENE
OLIGOCENE
EOCENE
PALEOCENE
CRETACEOUS
River alluvium and levees
Delta Plain
Chenier Plain
loess
Complexes of terraces and deposits
Fleming fm.
Catahoula Fm.
Vicksburg gp.
Jackson gp.
Claiborne gp.
Wilcox gp.
Midway gp.
only on Prothro and Rayburns salt domes
ROCK SYMBOLS
sandstone
siltstone
claystone
marl, calcareous clay
anhydrite, gypsum
limestone

Despite earth's advanced age, the rocks of Louisiana are remarkably young. The oldest rocks exposed at the surface are mudstones in the extreme northwest corner, which are about 60 to 65 million years old, laid down at the very beginning of Tertiary time, shortly after the demise of the dinosaurs. Sixty million years is an unimaginably long time in human terms, but just a little more than the last 1 percent of geologic time.

Louisiana is not so much the place to be continually impressed with the length of geologic time as the place to marvel at the rapid processes that built the state so quickly. Consider that the coastal plain has extended 250 miles into the Gulf of Mexico in deposits of sediment 50,000 to 60,000 feet thick in the time since the dinosaurs vanished. Louisiana's vast coastal plain has developed in just 7,500 years. Indians lived here before Louisiana's modern coastal plain had even begun. Nearly all the rocks you see along the roads were deposited during Tertiary time, the only exceptions a few patches of Cretaceous limestones that were pushed to the surface, through the overlying Tertiary rocks, by rising salt domes.

Geologic History of Louisiana

The story of Louisiana's geologic beginning is buried thousands of feet beneath the surface. It is known only from bits and pieces of rock brought to light by oil well drilling. Though not exposed in Louisiana, these early rocks appear at the surface in Texas, Arkansas, and Oklahoma, so the broad picture of what happened can be pieced together on a regional scale.

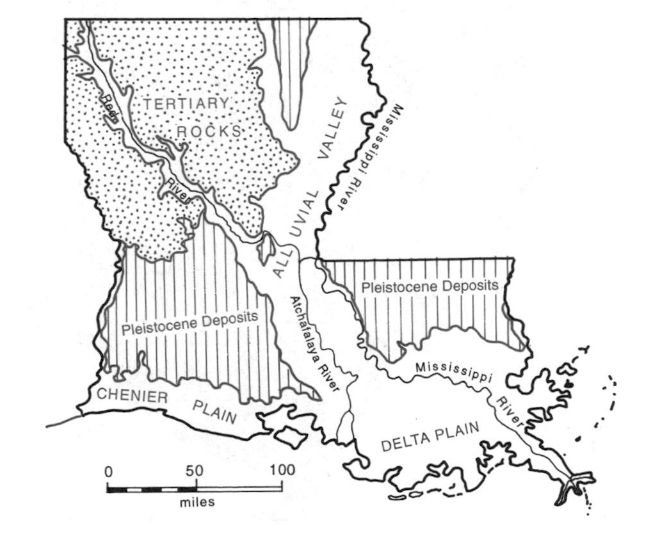

Basic geology of Louisiana.

The geologic tale of Louisiana is told entirely by sedimentary rocks, those deposited as grains or chemicals at the earth's surface. Most names of sedimentary rocks are easy to understand: sand becomes sandstone; mud turns into mudstone or shale; silt becomes siltstone; gravel turns into conglomerate; limy shells become limestone; salt is salt. You need no dictionary.

Geologists give names to distinctive packages of similar types of rocks, referring to them as formations. A cohesive bundle of formations may be given a group name. Louisiana's stack of sedimentary rocks, their group and formation names, and their respective ages are shown in the "Rocks of Louisiana" diagram.

The story starts about 250 million years ago, in Permian time, when the earth's continents were welded together in the great supercontinent, Pangaea. South America was then connected to North America along a curved line that starts in Texas, follows the southern border of the Ouachita Mountains through eastern Oklahoma and Arkansas, and continues to central Mississippi.

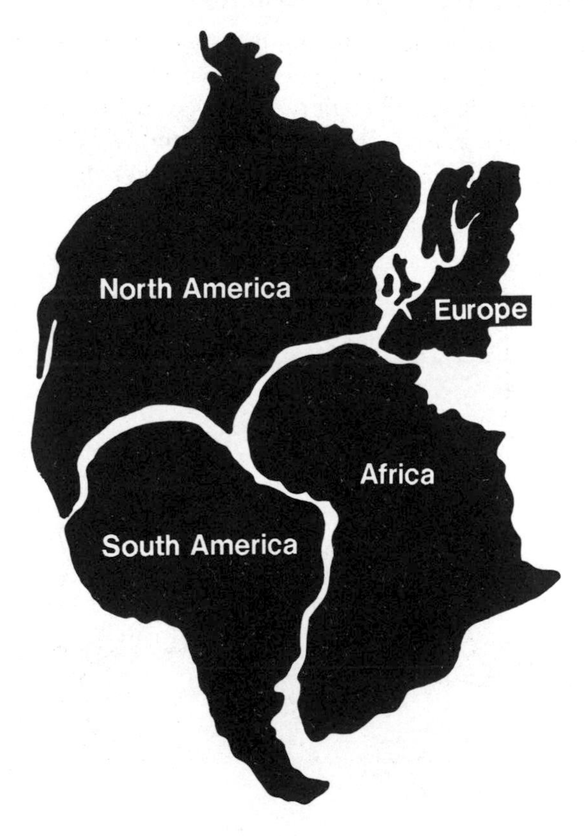

The supercontinent Pangaea, before breakup, showing how North America, South America, and Africa fit together.
—Wood and Walper, 1974

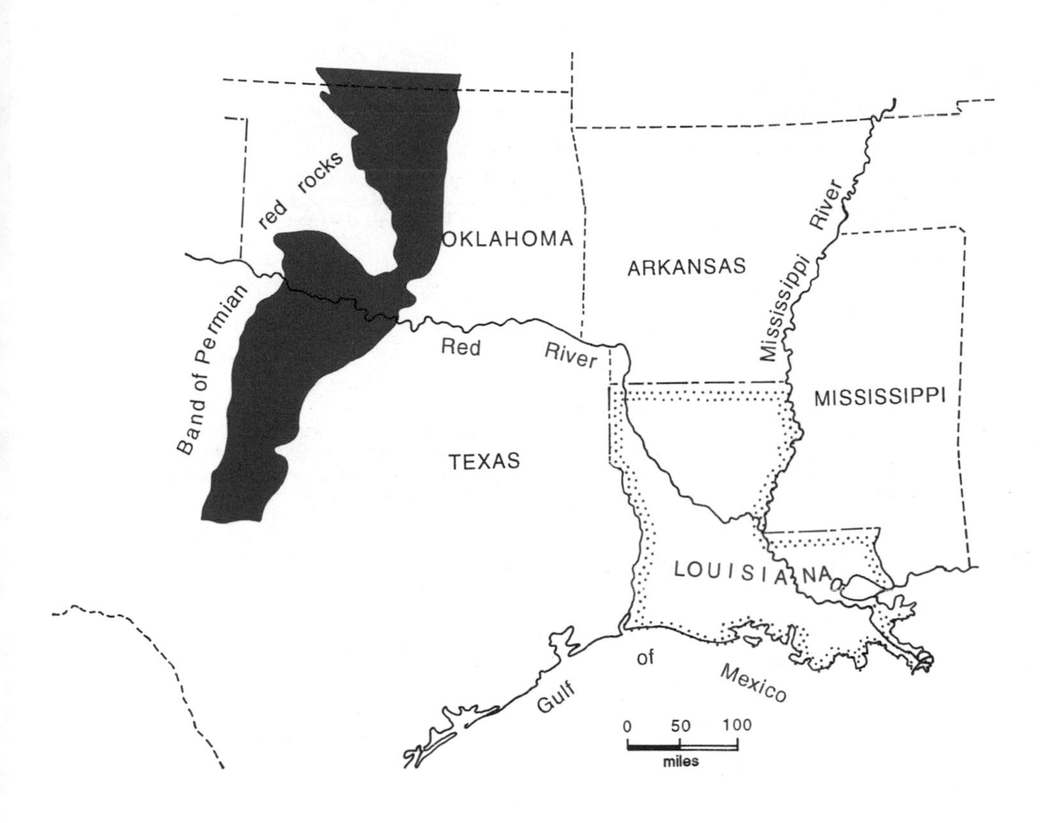

Band of red Permian rocks in Texas and Oklahoma is source of red sediment that gives the Red River its name. —Manning, 1990

Bright red sediments were shed westward into a sag called the Permian basin. Belts of these red beds form colorful hills that sweep across northern Texas and into Oklahoma. They are important to the geology of Louisiana because these Permian red beds donate the sediment that gives the Red River its distinctive hue and name.

Pangaea began to break apart in Triassic time, 245 to 208 million years ago. The earth's crust sagged along the line between what would later become North and South America, creating a depression that filled with shallow seawater, unconnected to the greater oceans, an early version of the Gulf of Mexico. It periodically dried up, leaving behind thick layers of minerals, such as salt, anhydrite, and gypsum. These salts accumulated to form the famous Louann salt, which today lies buried 20,000 to 50,000 feet beneath Louisiana. The Louann salt is the mother lode for Louisiana's numerous salt domes.

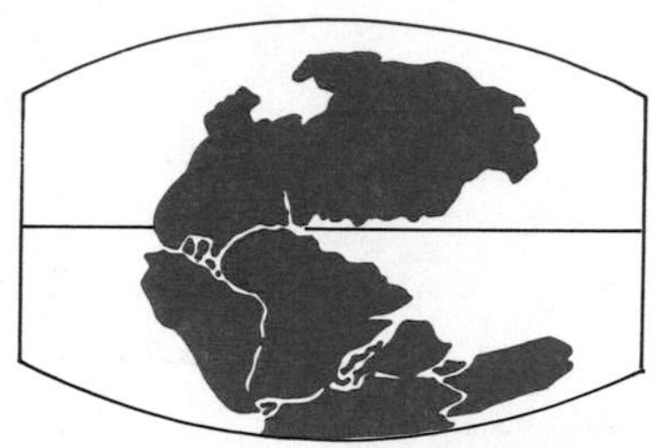

PERMIAN

245 million years ago
The continents were joined together in one supercontinent called Pangaea.

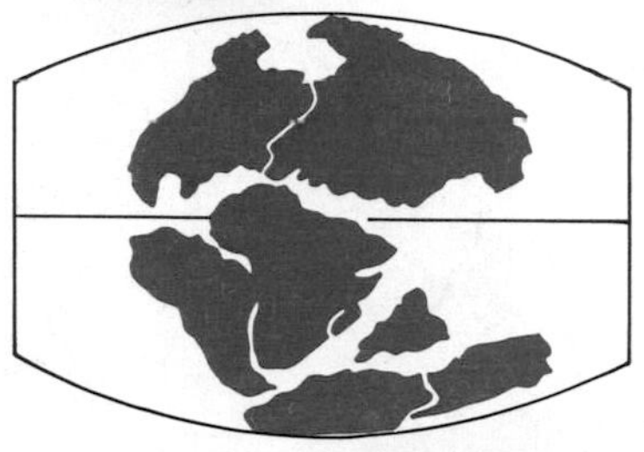

TRIASSIC

225 million years ago
Pangaea began to break apart as rifts formed from upwelling of lava from the hot, churning mantle.

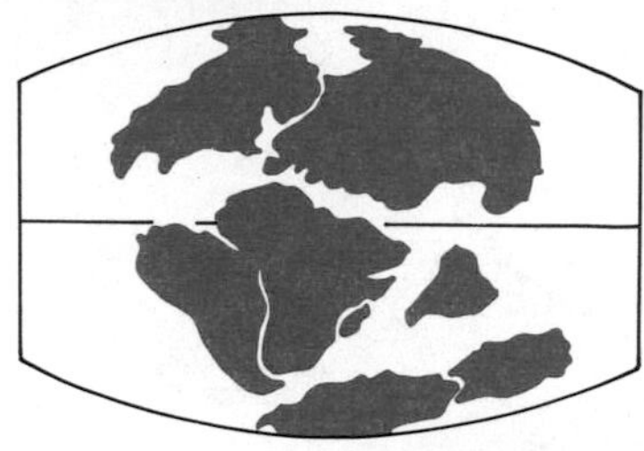

JURASSIC

145 million years ago
Continents continued to spread apart. South Atlantic begins to open.

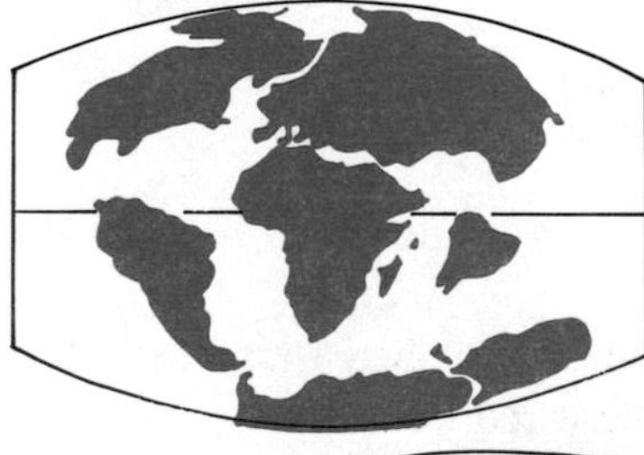

CRETACEOUS

66 million years ago
North Atlantic opens. India approaches Asia.

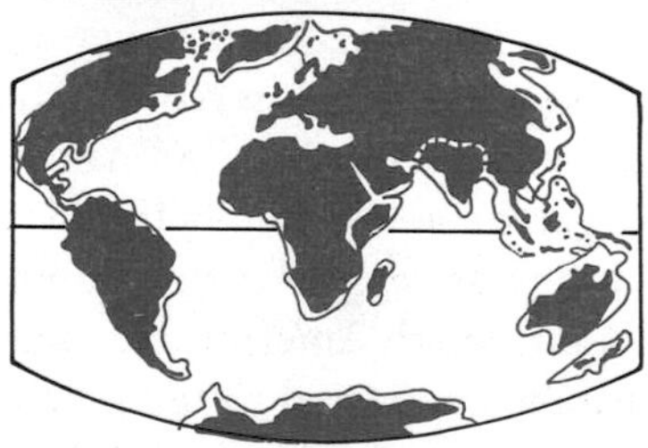

TODAY

Positions of the continents during the past 245 million years. —Spearing, 1991

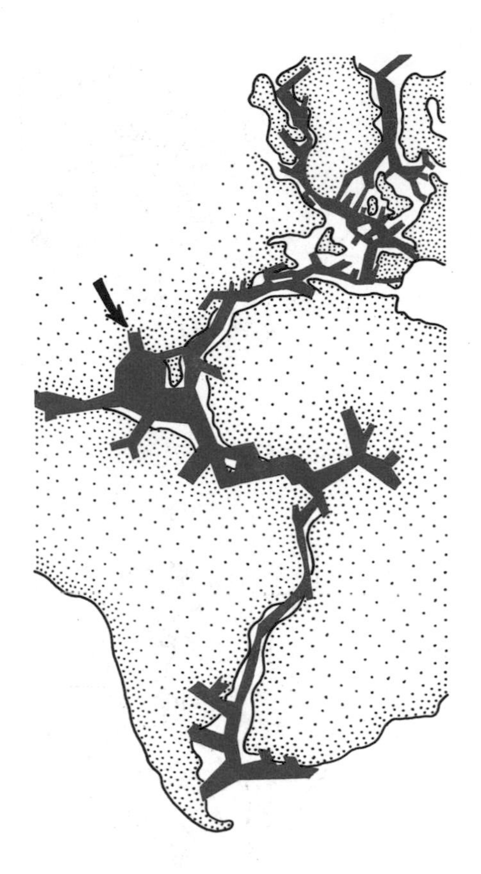

Ragged edges of continents are caused by small rifts that split off from the main rift. Arrow points to rift that set structural stage for the lower Mississippi Valley. —Burke, 1980

Louann salt deposition continued into the Jurassic period, when Pangaea really began to come apart. As North America moved west, the Rocky Mountains rose. A deeper Gulf of Mexico developed by late Jurassic time, as South America moved away from North America. Minerals accumulated on the drying flats along the northern edge of the Gulf of Mexico, shells of marine organisms piled up into carbonate layers to become limestone, and streams dumped their muddy sediments.

As the Cretaceous period of 144 to 66 million years ago unfolded, the Gulf of Mexico turned into a genuine ocean connected to the rest of the world ocean. The Rocky Mountains were growing rapidly, and much of the future Ouachita Mountains was buried under an accumulating pile of Cretaceous sediments. Sea level rose and fell many times during this time, driving water back and forth across the shallow shelf margin of the Gulf of Mexico and causing sandy river and shoreline deposits to alternate with shallow marine limestones. Meanwhile, marine clays and limy shales were deposited in the deeper waters.

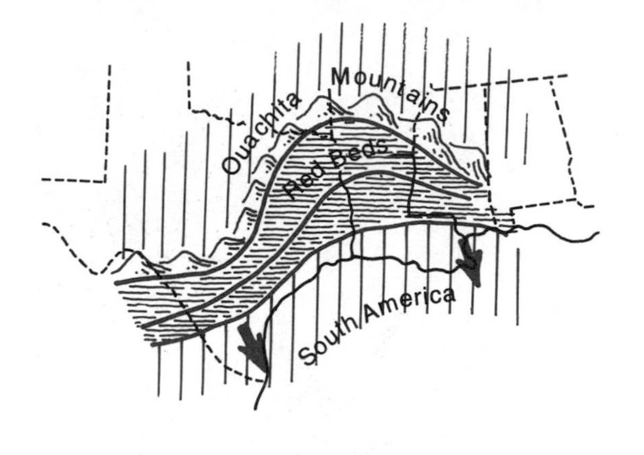

Triassic (245 to 208 million years).

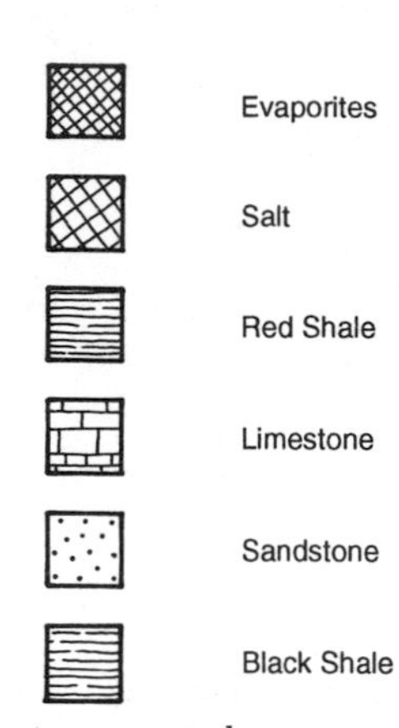

map key
—Burgess, 1976

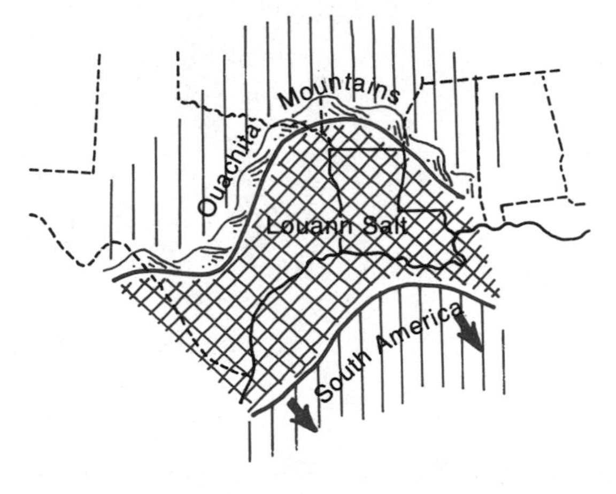

Triassic-Jurassic (about 215 to 200 million years).

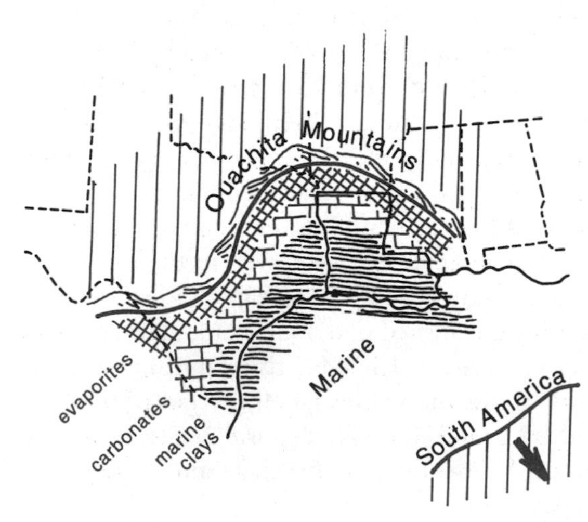

Jurassic (208 to 144 million years).

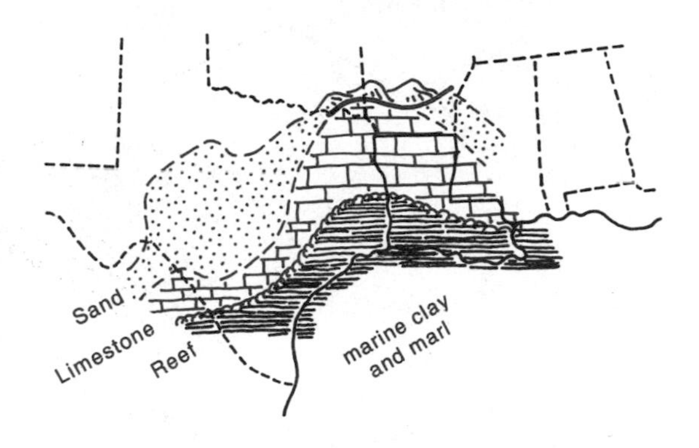

Cretaceous (144 to 66 million years). Most of the Ouachita Mountains are buried by this time.

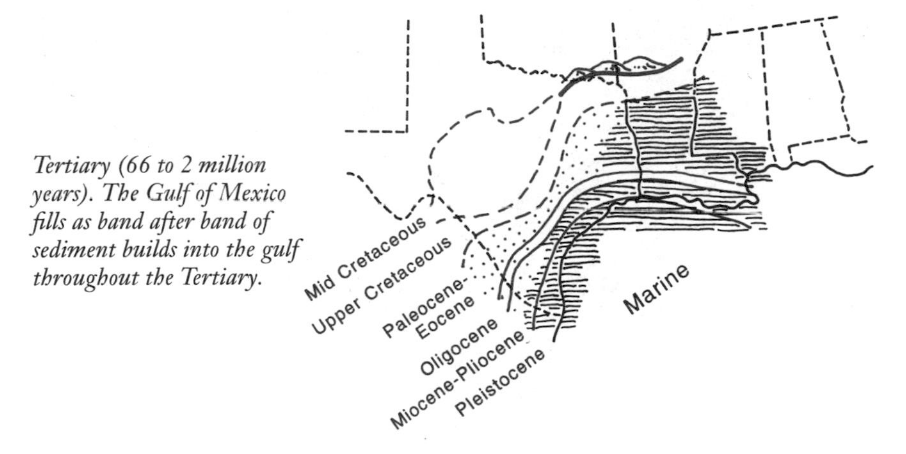

Tertiary (66 to 2 million years). The Gulf of Mexico fills as band after band of sediment builds into the gulf throughout the Tertiary.

This part of Louisiana's geologic history is known only from oil drilling. None of the rocks that record it are exposed at the surface in Louisiana, except in a few places where a rising salt dome pushed Cretaceous limestone to the surface.

Cretaceous time ended abruptly in a great cataclysm of extinctions that wiped out the dinosaurs, and many forms of marine life. Then, Tertiary time began.

Throughout Tertiary time, the environments in which sediments were deposited were remarkably similar to those we see today: river, floodplain, delta, shoreline, and shallow marine settings. Careful study of modern environments has helped geologists interpret the Tertiary environments by recognizing similarities between modern and ancient deposits.

The rocks you see in Louisiana accumulated as a pile of sediment that is still growing. Rivers poured a continuous supply of mud, silt, and sand onto the Mississippi delta and into the Gulf of Mexico. Giant wedges of sediment built into the Gulf of Mexico, tipping gently seaward, one piled on another in a progressively seaward direction. From north to south down the length of Louisiana the land grew gulfward, the oldest packages of

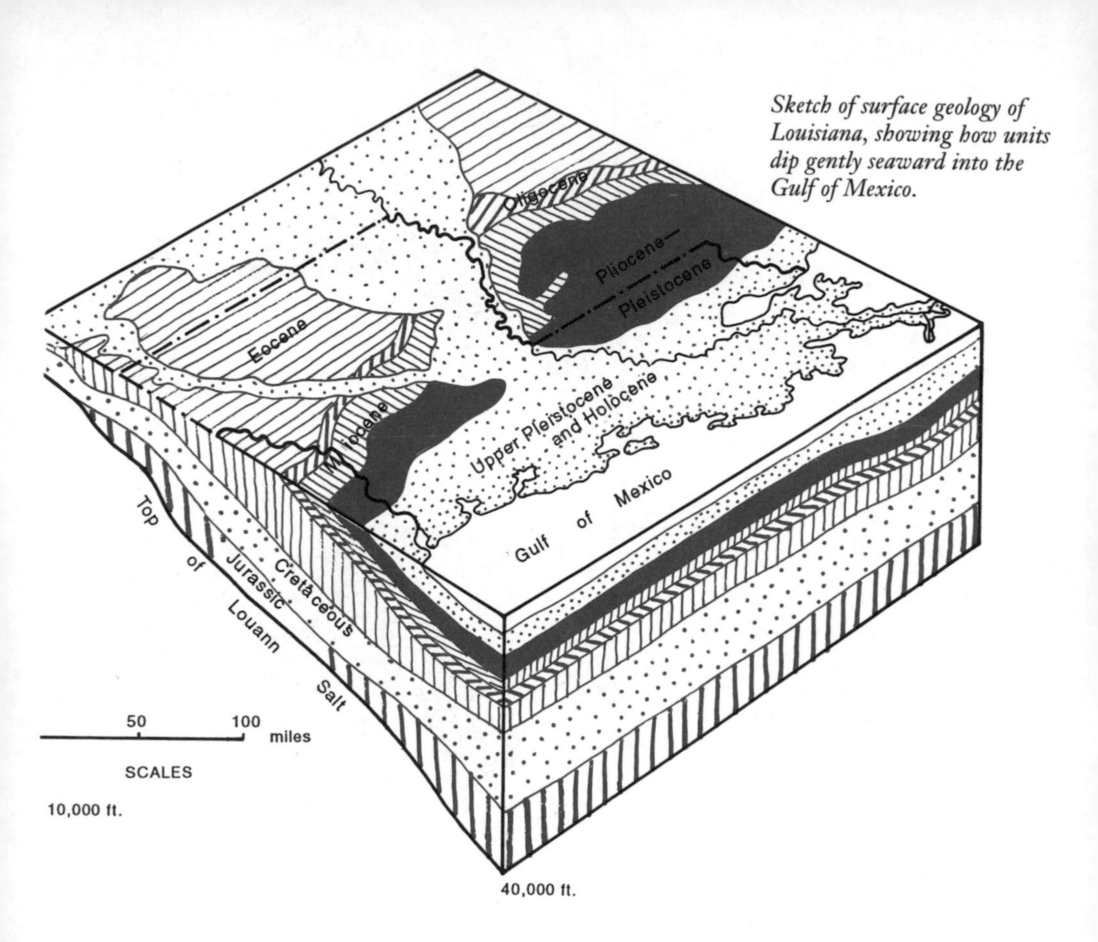

Sketch of surface geology of Louisiana, showing how units dip gently seaward into the Gulf of Mexico.

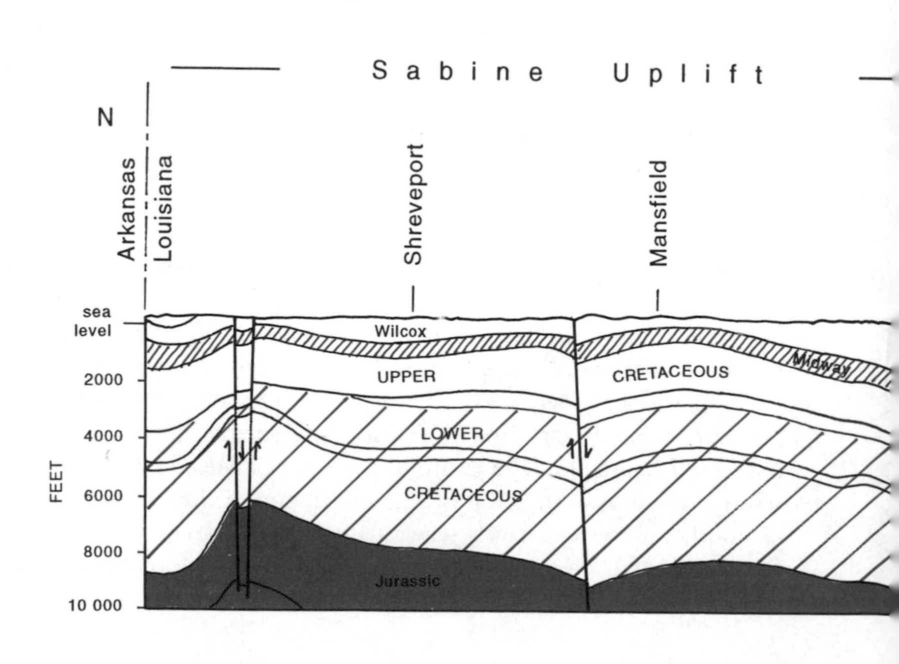

Above and below: *North-south profile* (from N to S) *across Louisiana. The arched Sabine uplift is prominent in northern Louisiana. Thickening wedges of Tertiary sediment dip into the Gulf of Mexico off the southern flank of the Sabine uplift.* —Bennison, 1975

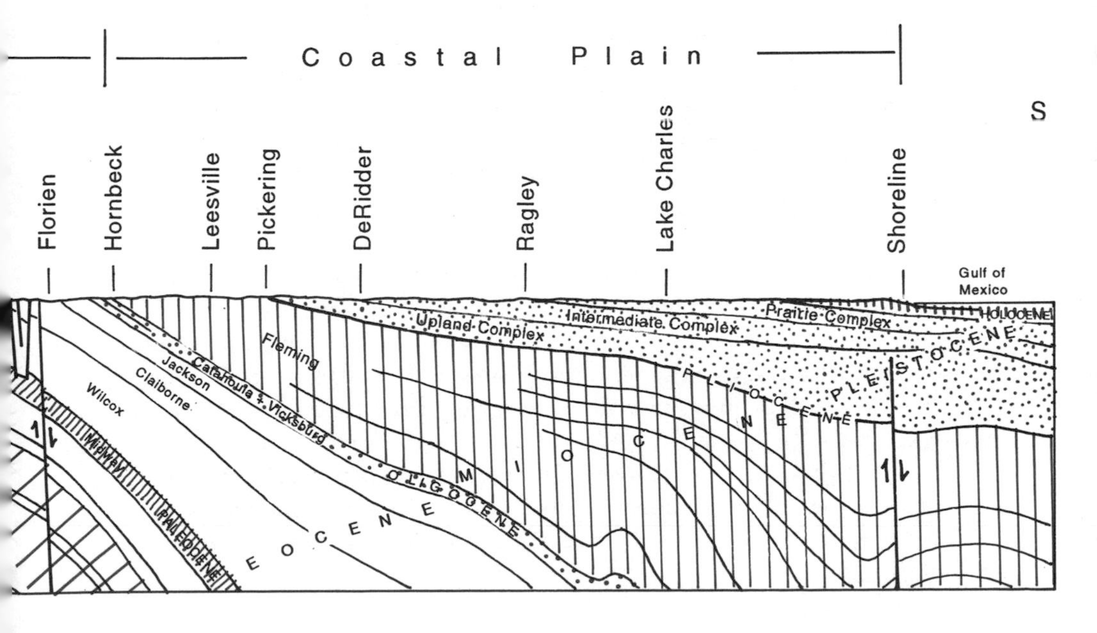

sediment in the north, the youngest in the south. These Tertiary sediments are most of the rocks you see in Louisiana. Their areas of outcrop make nearly parallel bands across the landscape and appear as a pattern of stripes on the geologic map. The roads of northwestern and central Louisiana ride over and cut through hilly displays of these Tertiary rocks. Many good roadcuts, quarries, and natural outcrops provide excellent opportunities to see them.

The oldest rocks in the state are a small patch of mudstones around Caddo Lake, northwest of Shreveport. They were laid down in Paleocene time, 66 to 55 million years ago, and are the only surface exposures of the Midway group of formations in Louisiana.

A broad area of Paleocene to Eocene rocks of the Wilcox group of formations exists in northwestern Louisiana, on the Sabine uplift, south of Shreveport. They include clays, sands, lignite coal, and silt, which vary greatly in color and extent. They are typical of the sediment that accumulates on a delta, and that is how geologists interpret them. Some of the lignite in the Wilcox group is mined to fuel electrical generating plants.

The Eocene Claiborne group overlies the Wilcox group in a broad band that sweeps around the eastern side of the Sabine uplift, across the North Louisiana salt basin. Sediments of the Claiborne group were deposited on a delta and in shallow seawater. The oldest formation is the Carrizo sandstone, a quartz sand with thin stringers of lignite coal. Above that is the Cane River formation, mostly clay with thin layers of silt and lignite; its fossils are the remains of animals that lived in shallow seawater. The Sparta sandstone, above the Cane River formation, consists mostly of sandstone, with layers of silt, clay, and lignite coal; the sandstone contains grains of a green mineral called glauconite, which forms only in shallow seawater. Next up is the Cook Mountain formation, which also contains glauconitic sand and clay; it weathers to a characteristic brownish red, largely an iron oxide stain derived from the glauconite, which contains iron. The Cockfield formation is at the top of the sequence; it consists of sand, silt, clay, and lignite, all deposited on a delta; it weathers to shades of light brown and tan.

The late Eocene Jackson and Oligocene Vicksburg groups form narrow bands of poorly exposed, mostly claystone deposits that lie south of the band of Claiborne group outcrops. Clays of the Jackson group contain some ironstone stringers and thin sandstones. The clays in the Vicksburg group were deposited in shallow seawater; they are limy and contain fossil seashells.

The Oligocene Catahoula sandstone is a fairly hard rock that supports ridges across central Louisiana. Its coarse sand was deposited in river channels. It has been called a rice grain sand because of its large grains.

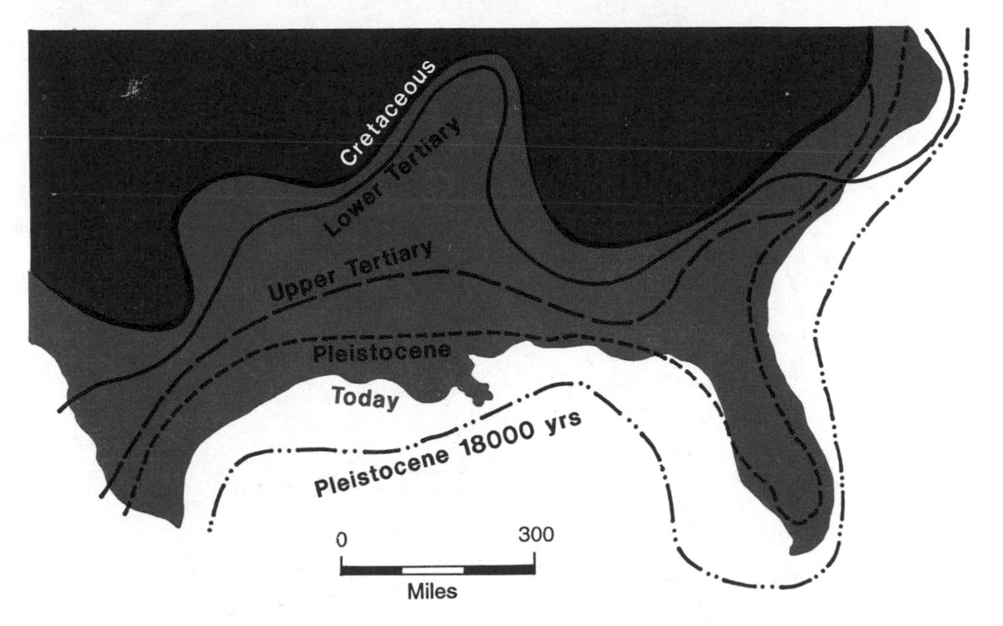

Shoreline positions along the Gulf of Mexico during the past 60 million years. —Walker and Coleman, 1987

The Catahoula sand contains volcanic ash, which apparently blew in from the west; it commonly contains petrified palm wood.

The Fleming formation is the youngest deposited during Tertiary time; it was laid down during Miocene and Pliocene time. The Fleming formation is a stack of clays in a variety of colors, thin layers of sand, volcanic ash, and siltstones, which all weathers to black soils. The sand and mud were originally river channel and floodplain deposits on which volcanic ash fell. These parts of the formation contain petrified palm wood, a few other parts the fossils of animals that lived in shallow seawater.

Nearly 2 million years ago North America shivered into the series of great ice ages of Pleistocene time. Time and again, the climate cooled and immense ice sheets blanketed much of Canada and the northern United States. These great glaciers profoundly affected Louisiana, even though none ever came closer than about 400 miles. When glaciers grew, they stored vast amounts of water, and sea level dropped dramatically, sometimes as much as 450 feet below its present stand. When the ice ages ended, the climate warmed quickly, and the glaciers melted quickly, sending monumental floods down rivers to the oceans. Sea level rose.

The coast of Louisiana was far south of its modern position when sea level was low during ice ages. Rivers entrenched their valleys as they adjusted to the lower sea level and dumped their sediment loads along that lower coast. In warm interglacial intervals, sea level stood higher and the

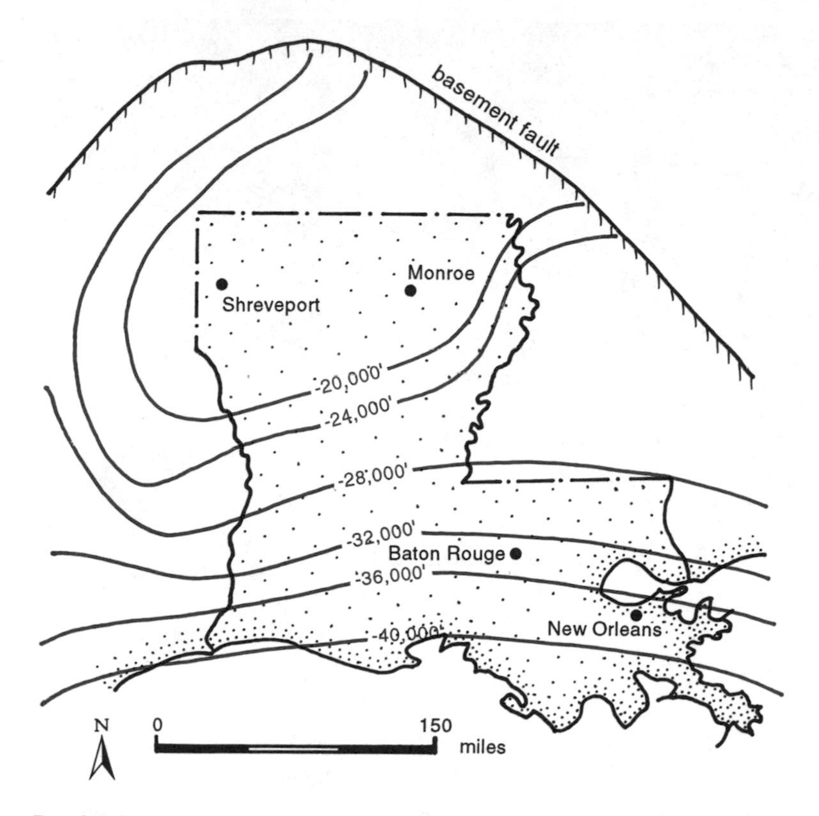

Depth to basement. Sedimentary rocks in Louisiana range in thickness from 20,000 feet in the north to over 40,000 feet thick along the coast. —USGS, AAPG, Basement Map of North America, 1967

coast was farther inland than today. Rivers filled their valleys with sand and gravel to adjust to the higher sea level.

About 120,000 to 130,000 years ago, during a particularly warm interglacial period, the coast was 35 miles inland from its present position. This old sandy shoreline passes through Sam Houston Jones State Park in western Louisiana, a few miles north of Lake Charles. Sea level dropped to an ice age low about 18,000 years ago, causing rivers to entrench and extend to a shoreline many miles seaward of the present coast.

Pleistocene sediments cover nearly a third of Louisiana. They sweep across the state in a broad arc from Lake Charles to Alexandria and across the Mississippi River to the Florida Parishes in three bands that span Pleistocene time from beginning to end. They consist mostly of gravel and coarse sand brought to the coast by rivers and deposited in great wedges that incline slightly seaward; they trend generally parallel to the modern coast.

The modern rivers have cut through the Pleistocene deposits, leaving remnants of them as terraces along the edges of valleys. They tell of higher river levels during interglacial times when sea level was higher. Macon Ridge, along the western side of the Mississippi Valley in northern Louisiana, is a deposit of gravel that an ice age version of the Mississippi River dumped during Pleistocene time.

The last great continental glacier was almost melted about 10,000 years ago, when sea level rose to nearly its present stand. It finally did reach that level about 7,500 years ago. Rivers partly filled their valleys with sediment as the rising sea flooded their mouths to create bays and lagoons. And the Mississippi River built the series of deltas that now make the modern coastal plain of wetlands, marshes, and swamps.

Structure Below

The sediments dumped into the Gulf of Mexico since it opened in Triassic time are a formidable pile. The stacked wedge rests on old rocks that were the edge of North America after it separated from South America. The pile of sediments is 20,000 feet thick in northern Louisiana, and thickens southward to 40,000 feet at the coast. It may be as much as 60,000 feet thick offshore.

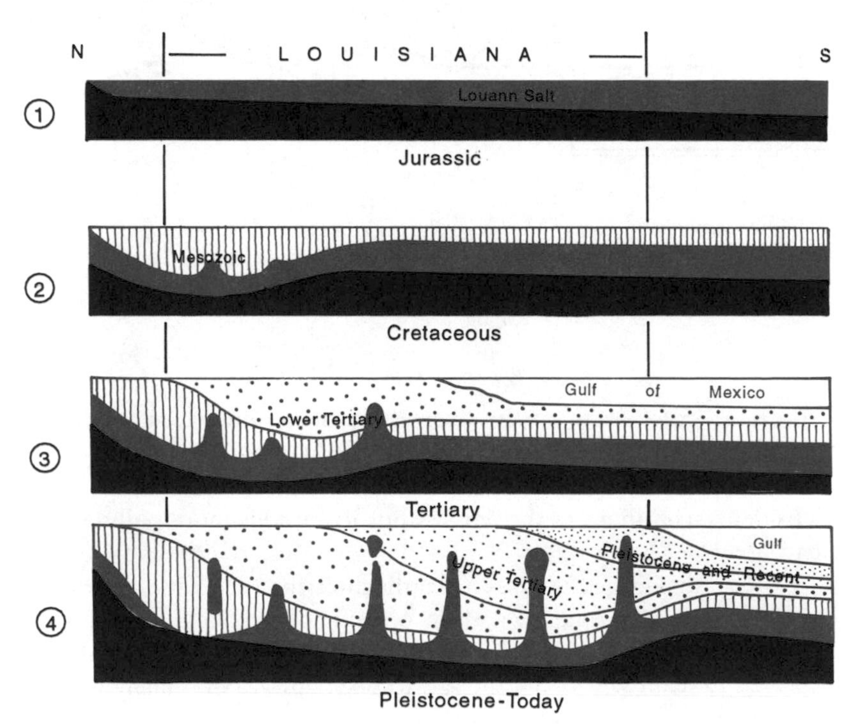

Stages in the development of the Gulf of Mexico, showing subsidence, progressive sediment fill, and diapiric salt movement. —Halbouty, 1979

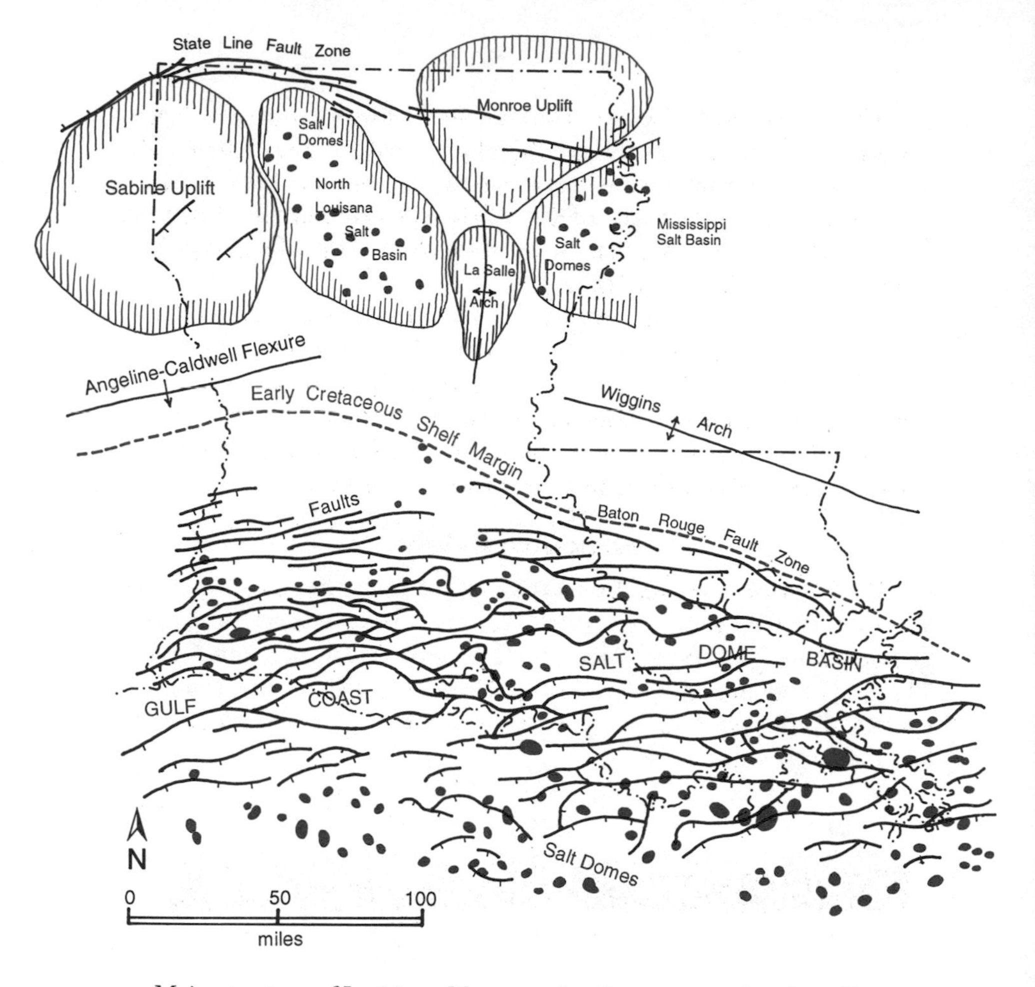

Major structures of Louisiana. Most are subsurface structures, but they affect many surface geologic features. Surface offset can be observed along the Baton Rouge fault, for example. —Salvador, 1991

The earth's crust floats on the deep rocks of its interior and sinks wherever a load is placed on it. Over time, the weight of sediments dumped into the Gulf of Mexico has pushed down the crust, further deepening the gulf. Subsidence continues as the Mississippi River adds more sediment every day.

Virtually all the deep structures beneath Louisiana reflect the separation of North and South America and the opening of the Gulf of Mexico. For example, the great weight of gulf sediments pushing down on the deeply buried Louann salt forced the salt to ooze upward into many vertical salt domes.

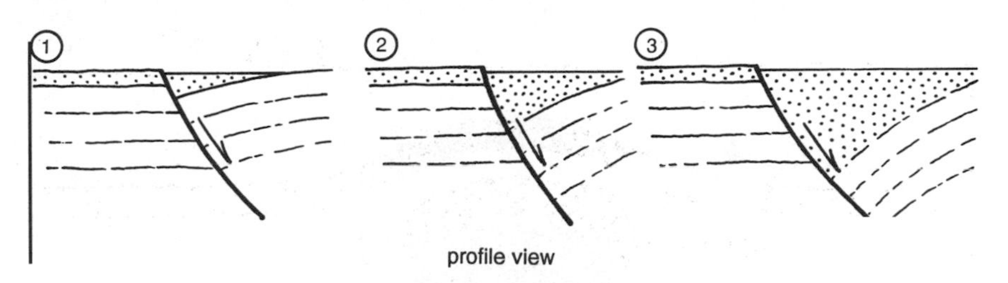

Continued slippage along growth fault causes sediment to fill in the low spot on the downthrown side. Thickened sand deposits along growth faults are the reservoirs for many oil and gas fields in the Gulf Coast.

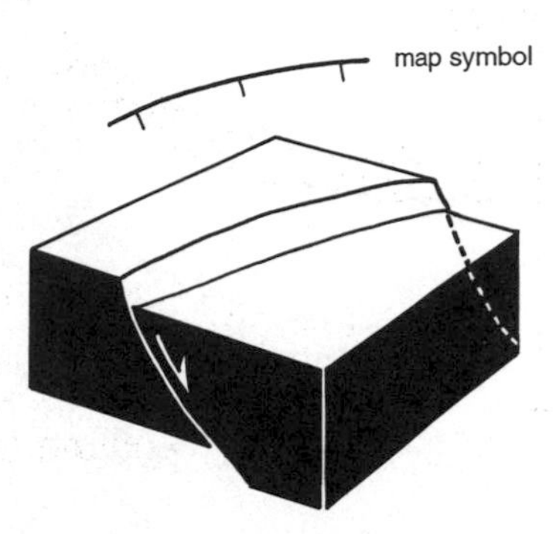

Diagram of a normal or growth fault common along the Gulf of Mexico (teeth on map symbol point to downdropped side).

Wherever an unusually large volume of salt has risen, the overlying sediments settle into compensating collapse structures. Such was the case in the Gulf Coast salt dome basin of southern Louisiana, which is in one of the thickest parts of the sediment pile. The collapse of the Gulf Coast salt dome basin is reflected in a host of sinuous faults that snake east to west through the sediment stack. The pile breaks apart along the faults as it slides southward toward the deeper center of the Gulf of Mexico. Folds and flexures that trend from east to west across central Louisiana also owe their existence to the collapsing sedimentary pile, as does a fault zone along the Arkansas border.

Salt Domes

The Louann salt, a thick layer of bedded salt, lies deep beneath Louisiana, under some ten miles of younger rocks. The salt was deposited in shallow depressions that opened between North and South America as they separated during Triassic and Jurassic time. Seawater periodically flooded the depressions, then evaporated, leaving behind layers of salt, gypsum, and anhydrite. Thousands of feet of Louann salt accumulated, then was buried under miles of sediment that the Mississippi River and its ancestors brought to the coast.

Salt is less dense than sandy and muddy sediments, so it rises to the top. We ordinarily think of salt as a brittle solid, but it flows like toothpaste

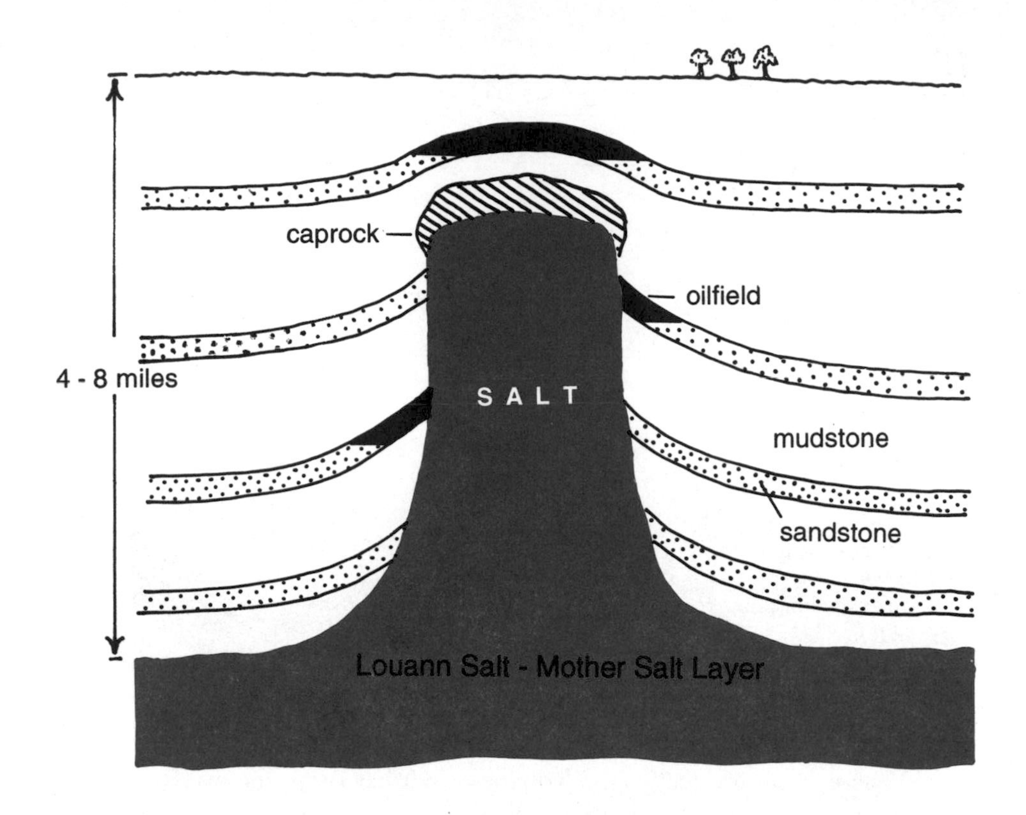

Cross section of a salt dome. The horizontal scale is exaggerated; salt domes are normally much taller and narrower than shown.

under the pressure of thousands of feet of rock. Salt also readily conducts heat, a property that may help heat flow from deep in the subsurface to the top of a salt dome, thereby allowing salt to flow at cool, shallow depths. The flowing salt rises mostly in vertical pillars, forming salt domes that rise from the depths. Some reach to the surface.

Oil and gas exploration in southern Louisiana reveals hundreds of salt domes, both onshore and offshore. Geologists call that area the Gulf Coast salt dome basin. More salt domes exist in the North Louisiana salt dome basin, near Shreveport. A few salt domes in northeastern Louisiana are in the western tip of the Mississippi salt dome basin.

Gulf Coast geologists have studied the mechanics of salt deformation for many years. They use the term diapir for a rising plug of salt driven upward by the weight of the overlying rocks. In a second process called downbuilding, the sediments above the salt collapse along curving faults, displacing the Louann salt downward. Small basins form, and undisturbed salt remains as high ridges and domes between the basins. The displaced salt below the basins moves down the continental slope.

Louisiana salt domes. —Louisiana Geological Survey, 1980

Louisiana salt dome basins. —Salvador, 1991

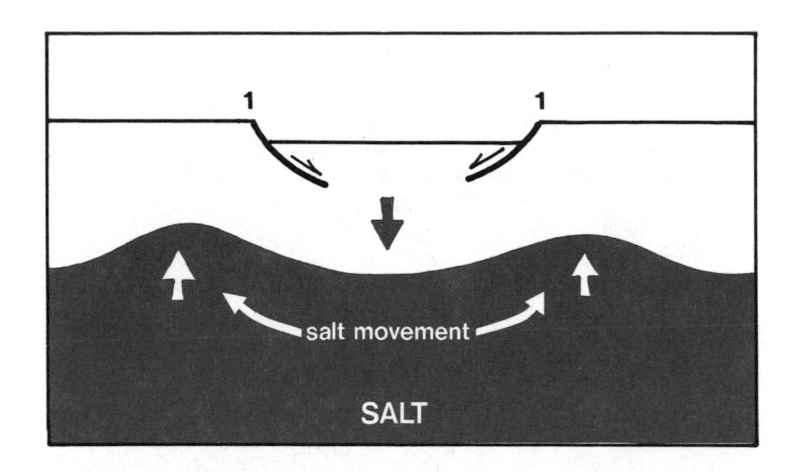

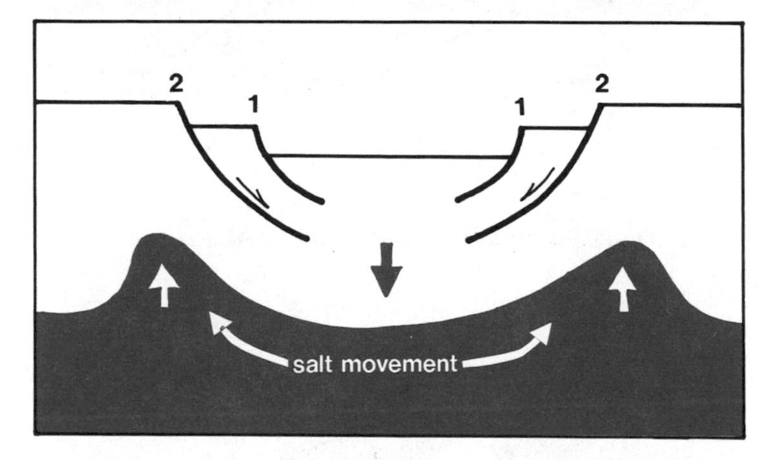

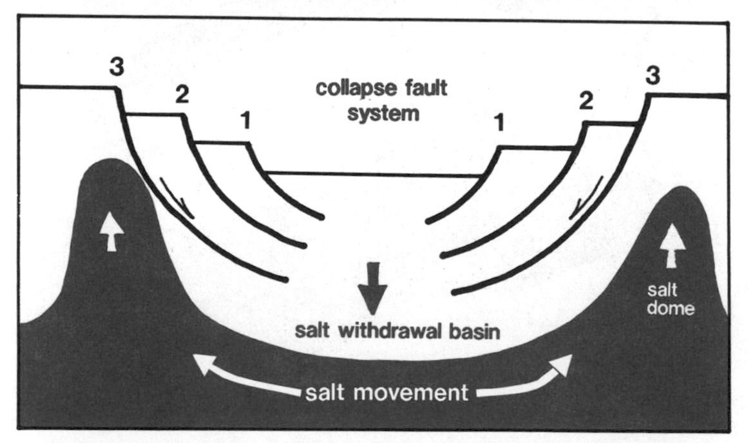

Collapse fault system. As the "mother" salt layer is pressed down by overlapping sediments, masses of salt move sideways and upward to form salt domes. The sediments collapse along faults into the space left by the withdrawn salt. —Seglund, 1974

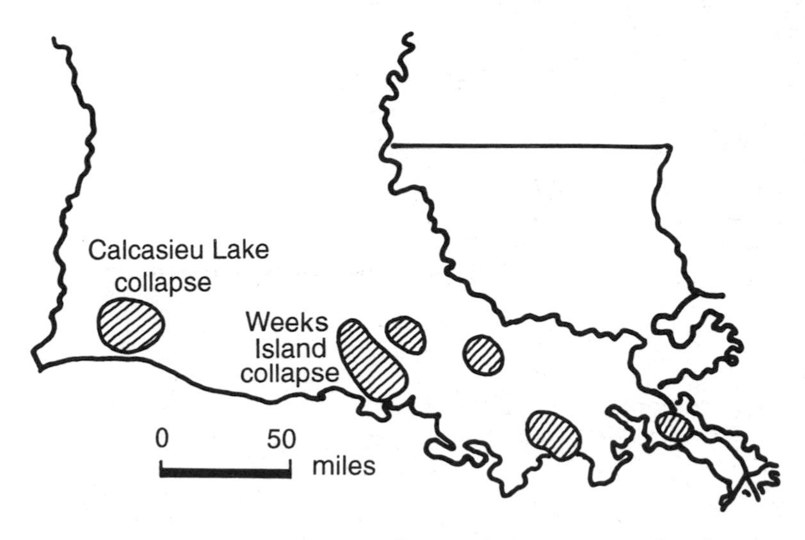

Collapse fault basins in the Gulf Coast salt dome province of southern Louisiana. —Seglund, 1974

In some parts of the Gulf Coast salt dome basin, collapse faults are associated with large salt domes or clusters of salt domes. These circular fault systems develop as sediments collapse above areas where salt has been withdrawn into a large salt dome or a cluster of salt domes. Four of the Five Island string of salt domes near New Iberia in southeastern Louisiana are associated with the Weeks Island collapse basin; the fifth and southernmost, Belle Isle, fortuitously aligns with the others. The Calcasieu Lake collapse basin underlies the marshlands near Hackberry, Gibsland, Cameron, and Creole, in southwestern Louisiana.

Much of the oil and gas in Louisiana are trapped in sandstone reservoirs tipped up against the impervious walls of salt domes. Oil and gas are also trapped in the arched and faulted sediments above salt domes. Brine for chemical processing is extracted from man-made caverns in salt, and crude oil is stored in underground salt caverns by the U.S. Department of Energy to maintain national petroleum reserves.

Caprock forms on top of salt domes as the domes rise and encounter groundwater, which dissolves the salt. As water dissolves holes in the salt, insoluble residues of calcite, gypsum, and anhydrite remain behind. The caprock contains no oxygen, so anaerobic bacteria thrive. They include sulfate-reducing bacteria, which attack the calcium sulfate minerals gypsum and anhydrite. The bacteria break the sulfate down to pure sulphur and sulfide, which combine with metals in groundwater to make a variety of metallic sulfide minerals. Sulphur is mined from several salt domes using steam to melt the sulphur and drive it to the surface as a liquid.

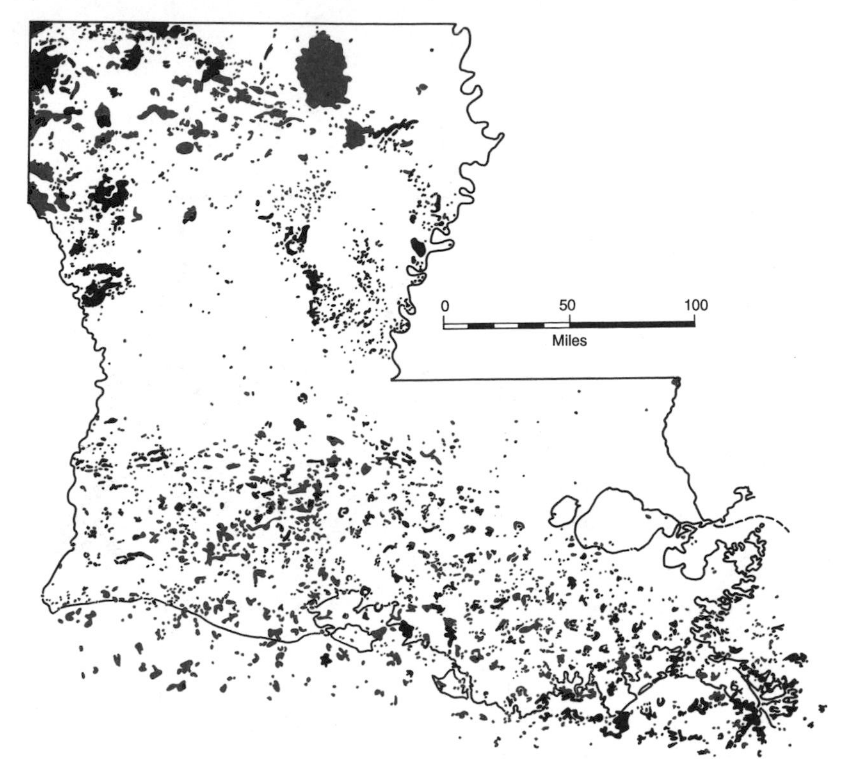

Oil (black) *and gas* (red) *fields of Louisiana.*
—Louisiana Geological Survey, 1981

Salt domes were probably the first geologic features to be remotely sensed in oil exploration. Because salt is less dense than most sedimentary rocks, salt domes could be located by carefully measuring small gravity variations, then drilling where the gravity measured low. Many of Louisiana's salt dome oil fields were found this way.

Most salt domes are hidden beneath the surface, but several mounds stand 80 to 100 feet above the surrounding terrain. They look oddly out of place in the level marshes and swampy countryside of coastal Louisiana. The Five Island salt domes are the most impressive. They stand in a line west of US 90 between New Iberia and Morgan City. From north to south, Jefferson, Avery, Weeks, Cote Blanche, and Belle Isle "islands" are forested hills surrounded by marsh and low ground.

Several domes in the North Louisiana salt basin approach the surface to form mounds southeast of Shreveport. The most interesting of these is Winnfield salt dome on US 84, between Winnfield and Natchitoches (pronounced *NACK-o-tish*). The road climbs the flank of the dome, providing visitors with a terrific view into the caprock that's exposed in the walls of the Winn Rock quarry.

Oil and Gas

Deep beneath Louisiana is a gigantic natural factory that has been operating for millions of years, churning out its black liquid product, crude oil—and a lot of natural gas. The crude oil factory was built by the natural geologic forces peculiar to Louisiana, within about the past 65 million years.

Rivers have brought sand and mud to the Gulf Coast since the end of Cretaceous time, 65 million years ago. Sand is deposited in long strings in channels and natural river levees, in piles at the river mouths, and in linear barrier islands along the shores. The mud spreads over the riverbanks during floods, to be deposited on the broad surfaces of deltas, which build sideways as well as seaward. The delta sand and mud are full of nutrients; the waters off delta fronts teem with life. The swamps and marshes, too, are incredibly rich in organisms. As the plants and animals die, their remains mix with the surrounding sediment to be preserved and buried.

Finally the rich organic sediment soup reaches burial depth of a few thousand feet. The temperature is hot. The plant and animal debris cooks, and tiny droplets of oil emerge. Meanwhile, the weight of the thick sediment pile compresses the deeper layers, jamming the sediment grains

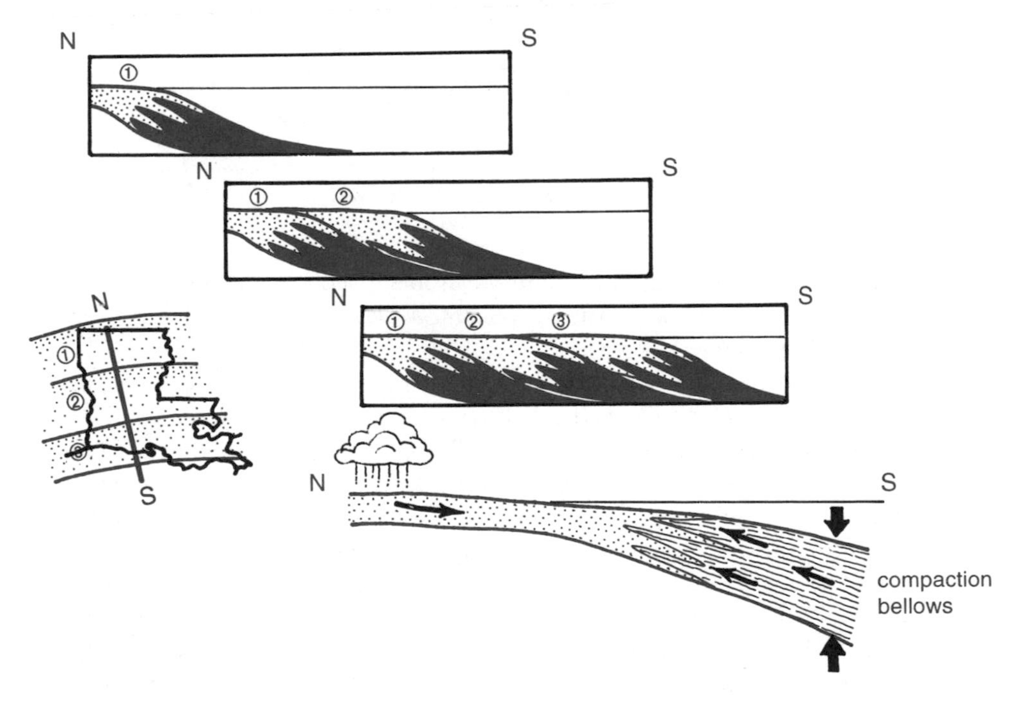

Louisiana deep magic! The sediment wedges filling the Gulf of Mexico place sandy river deposits (stippled) *against organic-rich marine mud* (red). *Thickening and burial of the wedges cause compaction, which acts as a "bellows," squeezing out water and oil from the muds (source rock). Water and oil migrate up to fill holes in the sands (reservoirs).* —Roberts, 1982

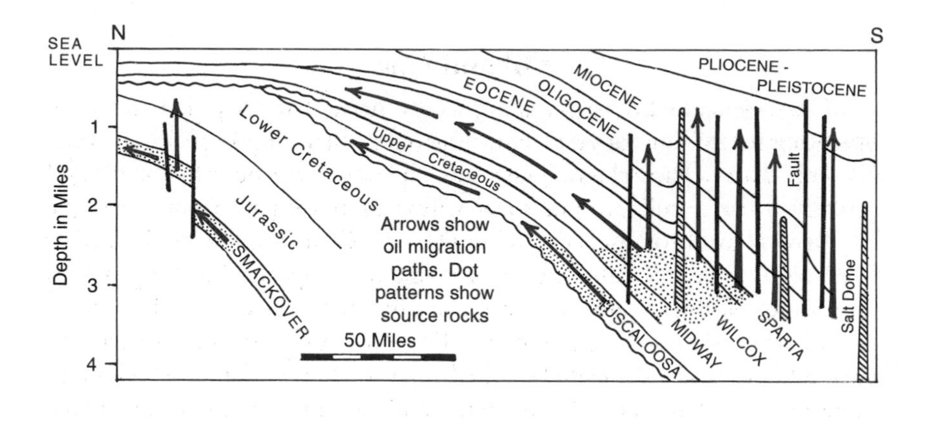

Principal reservoirs in the productive oil and gas belts across Louisiana. Faults and salt domes provide avenues for migrating oil. —Sassen, 1990

together, forcing out the water and the droplets of oil. Water and oil move up through the pile of sediment, concentrating along faults and along the edges of salt domes.

The lighter oil rises through the water and fills spaces between grains in the sandstones. It fills the buried stringers of river sand, the old barrier islands, the buried sand shoals, and sandbars. And the water seals it in from beneath. If conditions are right, the oil will remain stored in these sandy reservoirs for millions of years, until someone drills a well and sucks the oil to the surface to power a Chevy, light a home, or heat a school.

The intertwined environments we see today in Louisiana—river, delta, swamp, marsh, and barrier island—operate in concert over a very long time. Together they create the special elements of source rocks, cooking conditions, migration, reservoir, and trap that all must come together at the right time and place for oil to form and then find its way into reservoir sandstones in commercial quantities.

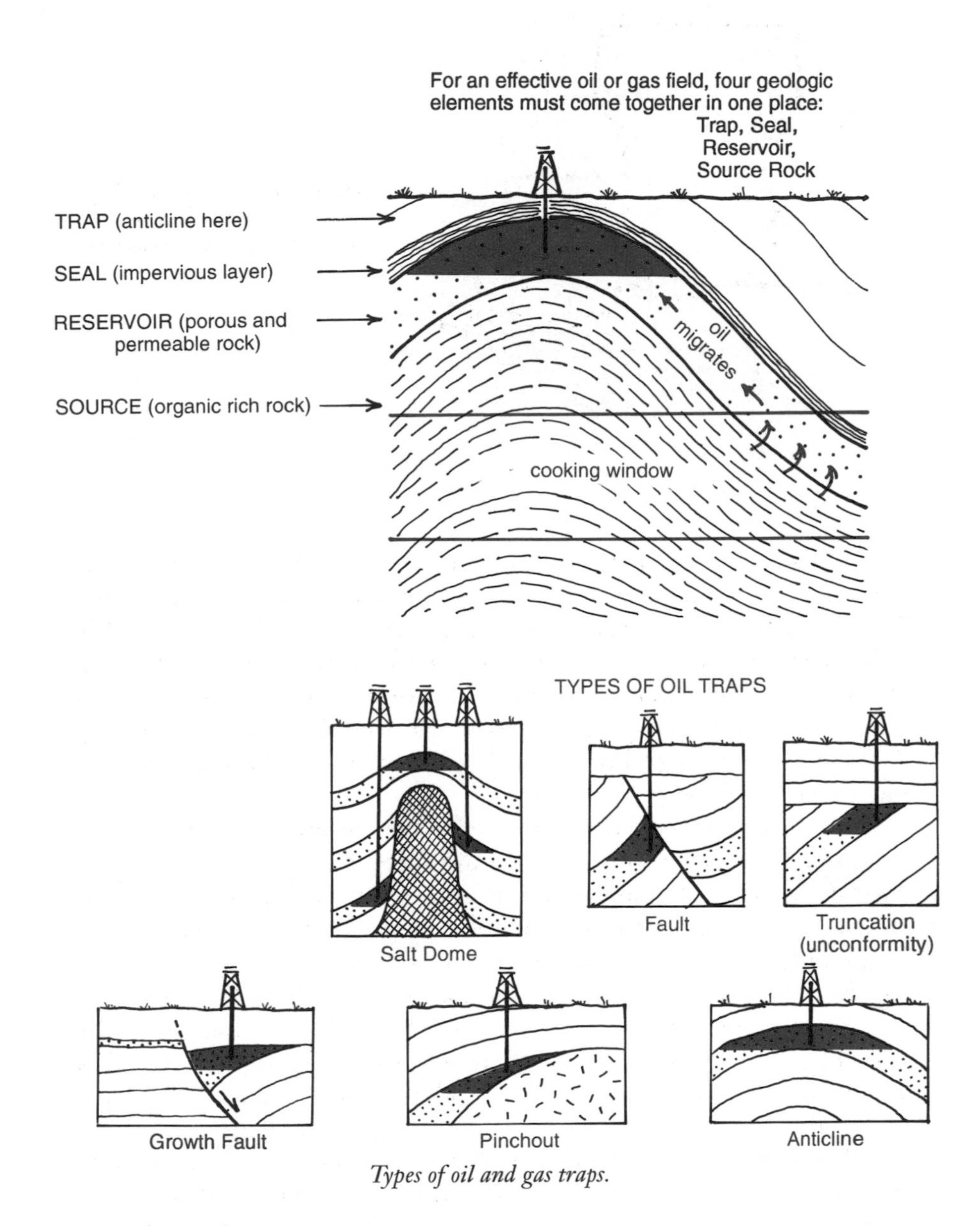

Types of oil and gas traps.

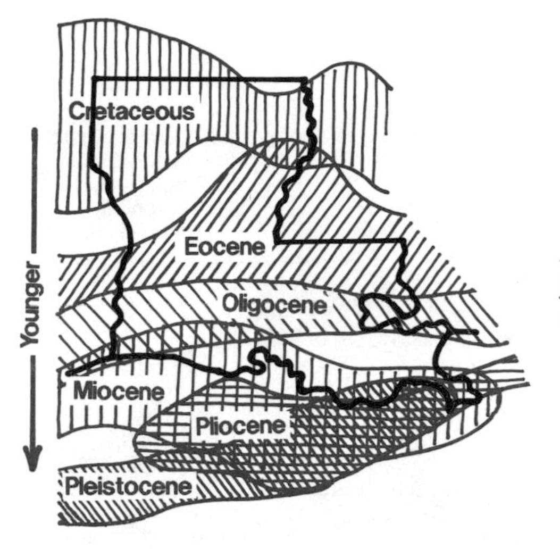

Ages of the reservoirs in the productive belts of oil and gas across Louisiana. —Roberts, 1982

Louisiana is one of the world's great oil and gas regions, with an annual production that ranks third in the United States, after Alaska and Texas. The history of Louisiana petroleum started in 1896, when Belle Isle salt dome was drilled, one of the first such ventures on the Gulf Coast. Although it did not produce commercial quantities of oil, the well did help inspire the giant Spindletop salt dome discovery in East Texas in early 1901. It also influenced Louisiana's first commercial oil field discovery on the Jennings salt dome in late 1901.

The Caddo gas field was discovered in the northwestern corner of the state in 1905—the first important gas discovery in the northern Louisiana district. According to some people, it was also the first offshore discovery, because the wells were drilled from wooden platforms in Caddo Lake. The bulge of the Sabine uplift played a vital role in generating oil traps in northwestern Louisiana. Parts of the Caddo field still produce today.

The Creole oil field—Louisiana's first truly offshore discovery—was found in open waters of the Gulf of Mexico in 1937 and 1938. The world's first offshore oil field out of sight of land was discovered in 1947, with a well drilled 12 miles from the shoreline of Terrebone Parish.

About 95 percent of Louisiana's petroleum comes from the southern part of the state, where numerous salt structures and rich deltaic source rocks combine to make a first-class oil and gas province. Northwestern Louisiana produces petroleum from traps around the Sabine uplift. The North Louisiana salt dome basin holds substantial reserves. Gas has been found in abundance around the Monroe uplift in northeastern Louisiana.

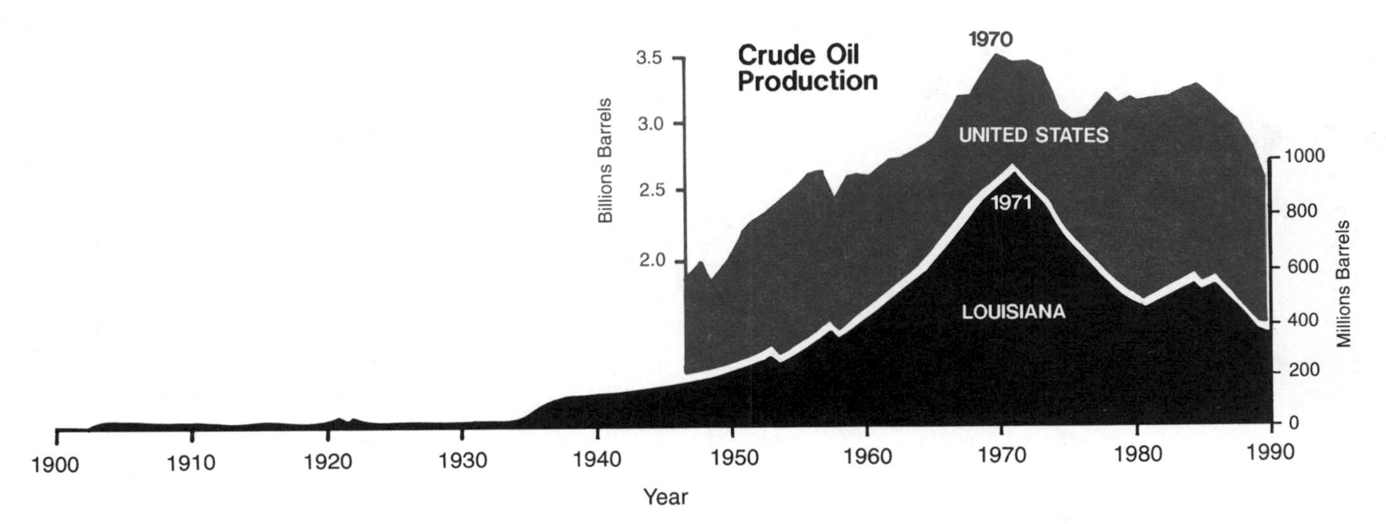

History of Louisiana's crude oil production since its beginning in 1902. Today, Louisiana is the nation's third largest producer behind Texas and Alaska. —compiled from U. S. Dept. Interior, Bur. Mines, Mineral Indust. Surveys. and U. S. Dept. Energy, Petrol. Supply Annuals

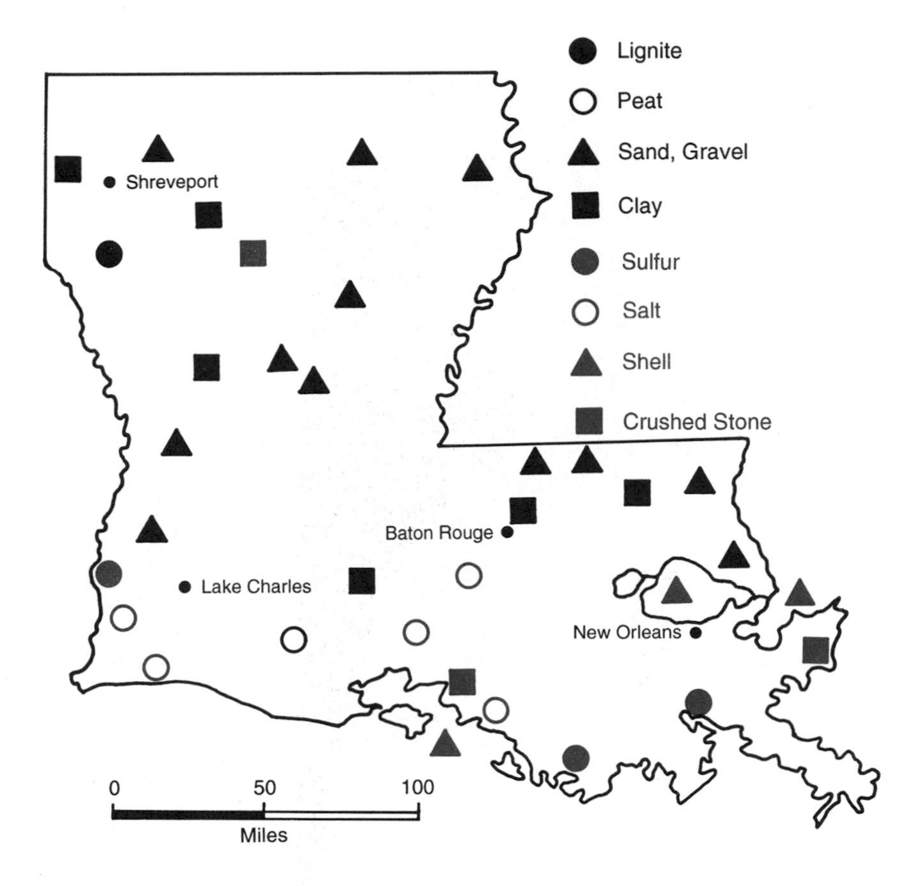

Mineral production (other than oil and gas) in Louisiana. —Autin and John, 1990

At its peak in 1971, Louisiana produced about 950 million barrels of oil, about a quarter of the United States' production. Since then, production in both Louisiana and the nation have steadily declined. Louisiana now produces about 300 to 400 million barrels of oil annually.

Minerals

Lacking mountains, igneous rocks, metamorphic rocks, hard rocks, and really old rocks, Louisiana is not the place to spend a fortune searching for gold, silver, or other hard rock minerals. But a surprisingly steady stream of rock materials flows from the ground to meet the state's needs.

Salt domes in southeastern and southwestern Louisiana yield sulphur from caprock layers to feed an enormous chemical industry that makes sulphuric acid, tires, and a host of other products. Crushed rock, salt, and salt brine are also extracted from salt domes. Hard caprock, mainly anhydrite and limestone, is highly prized in a state lacking hard materials. A huge caprock quarry near Winnfield supplies crushed rock for construction and road building. A large part of Louisiana's salt is used in the northern states to clear snow and ice from roads.

Ancient stream deposits, as well as modern river channels and floodplains, are mined for sand and gravel to use as construction aggregate. Shells are also used as low-cost gravel in roads. A road shoulder made of shell naturally compacts into a hard, limy surface that inhibits plant growth—important in Louisiana's semitropics, where vegetation quickly becomes junglelike. A local shell-dredging industry, both offshore and in Lake Pontchartrain, once met some of this market, but adverse environmental effects have sharply curtailed the dredging. Clean silica stream sand for making glass and for sandblasting is quarried from the Eocene Sparta formation in northwestern Louisiana and from Pleistocene deposits in southern Louisiana.

Clay mined from Pleistocene layers and from Mississippi River alluvium in southern Louisiana becomes ceramics, bricks, tile, and oil-drilling mud. It also finds its way into chemical manufacturing.

The rocks of the Wilcox formation in northern Louisiana, originally deposited in rich delta swamps during Paleocene time, contained vegetation that, when buried, became peat and later became lignite coal. The lignite fuels electricity-generating plants. Peat is dug in southwestern Louisiana for use as a soil additive.

Rivers in Motion

Rivers are in constant motion: eroding here, flooding there, twisting and turning in their valleys, even slicing through bedrock. You can stand on the bank and watch the water pass, but you will see the river in action only during a great flood. Only then does the river reach its potential in carrying sediment, eroding its channel, and ignoring natural levees, dikes, and other restraints.

A river moves like a snake, if you could view it in fast time. Its channel wiggles back and forth in loops and turns, lashing sideways and downriver through the soft sediments of its valley. This meandering motion is especially characteristic of rivers in low areas, such as Louisiana. But why does a river form meander loops? Why does it not flow in a straight line to the sea? The answer lies in natural irregularities.

Even if a river were originally straight, it would soon meander. Any irregularity, such as a caved bank, fallen tree, or the entry of a tributary stream, would deflect its flow toward the opposite bank, which it would erode. The eroded area would then deflect the flow to the other bank,

Louisiana rivers.

which the stream erodes next. And so the original deflection would create a series of bends on down the stream. As the stream continues to erode the outside of its bends, the bends grow into meanders.

The river erodes sediment from the outside banks of its bends, then deposits it in the slack water on the inside bank of the next bend downstream. That deposit, called a point bar, also helps deflect the flow, further exaggerating the meander loops. When the meanders become nearly circular loops, the river shortcuts across the narrow neck of the loop during a flood. Then the river abandons the severed loop, leaving it as a curving lake on the floodplain, an oxbow lake, which is common in Louisiana.

Everyone is familiar with the term floodplain. Floods occur naturally when the drainage basin dumps more water into the stream than its

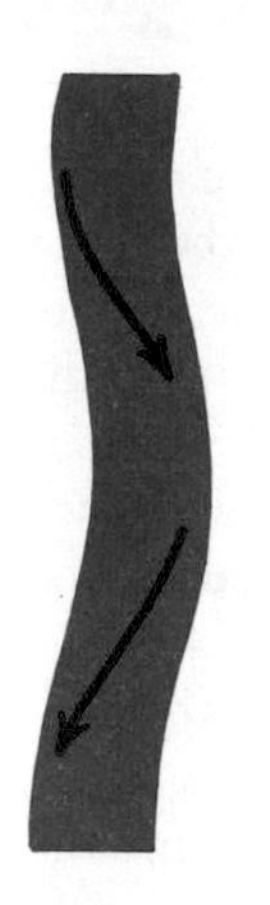

Straight river channel erodes unevenly due to uneven flow and irregularities.

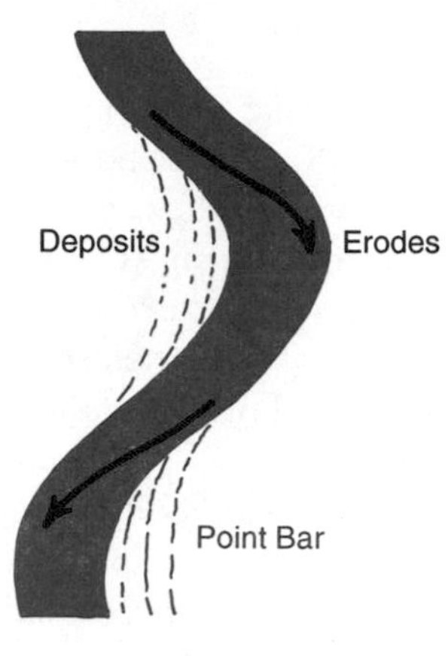

River erodes at outside bends, while sediment is deposited in slow water at inside of loops.

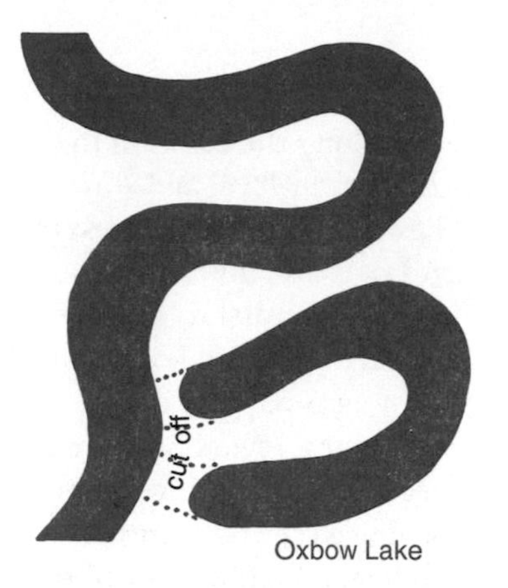

River cuts off narrow neck of a meander loop creating an oxbow lake.

channel can carry. Water rises out of the channel and spreads across the floodplain, where it stays until the channel can drain it away. The floodplain is where the river temporarily stores its excess water. One effect of protection levees is to deprive the river of part of its temporary water storage, which forces it to pile the water deeper in the parts still at its disposal. So a levee in one place may cause higher floods in other places.

Floodplains are flat because floods spread sediment across them, filling in the low places. The water slows as soon as it leaves the channel and overflows onto the floodplain. This happens along the stream banks, so that is where the water deposits a large proportion of its sediment load.

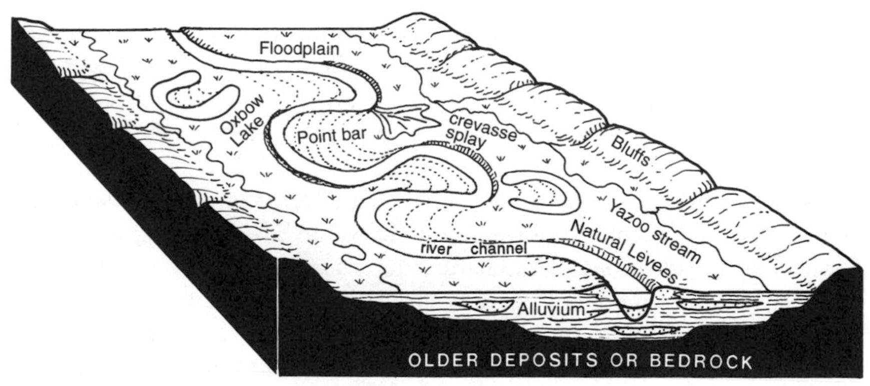

River floodplain environments and deposits.

This process builds the banks higher, creating a natural levee. The natural levees along the banks of the Mississippi River rise about 15 feet above the normal river level and slope gently away from the river for a mile or two. They are the highest, driest parts of the floodplain—major rivers in southern Louisiana flow on the highest land, not the lowest! Tributary streams may flow parallel to a major river for many miles before they find a place to cross the natural levee and join the river. These lateral streams are called rim-swamp bayous, or yazoo streams.

Natural levees help contain many average floods within the channel. Large floods overflow them infrequently, though some breach the levee, opening a crevasse through which water dumps a fan of sediment onto the floodplain. Geologists call those fans crevasse splay deposits. Some levees consist nearly entirely of coalesced crevasse splay deposits. The river repairs some crevasses with new deposits of sediment; others may become the point of origin of a distributary stream, and a few divert the entire stream onto the floodplain, leaving the old channel behind as a bayou.

Floodplains are typically rich farmland periodically replenished with fresh deposits of sediment. Some floodplains accumulate partially decayed vegetation as deposits of peat, which eventually turns into coal if younger sediments bury it. Lignite coal deposited on an ancient floodplain in the Paleocene-Eocene Wilcox group is close to the surface near Mansfield, west of the Red River in northwestern Louisiana. It is mined and burned to generate electricity.

Many sizable rivers in southern Louisiana are called bayous. A bayou may be a lazy swamp channel or a seasonal tributary of a river, but it is not necessarily small. In Louisiana, any watercourse can be called a bayou, from a swift hill-country creek to a former course of the largest river on the continent.

Dynamic Deltas

As a river builds its delta into the sea, it creates the conditions that will later force it to turn elsewhere, to build another delta. The river deposits sediment where it enters the sea, building the delta in two ways.

First, the water flows into the sea as a continuing stream, much as water streams out of a garden hose. The flow speed slows at the edge of the stream, causing sediment to settle. This effect creates an underwater natural levee on each side of the stream of water. The submerged natural levees are continuations of the natural levees you see above water. They grow until they emerge above the surface, at the same time extending seaward. Once the levees are in place, they confine the river, except during large floods. Floods add more sediment to the natural levees, and dump sediment into the shallow seawater beyond them. This fills the bays, in-

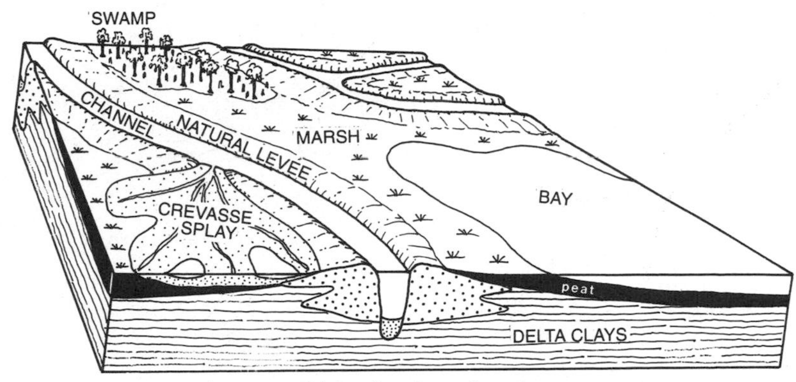

Anatomy of delta deposits and environments.

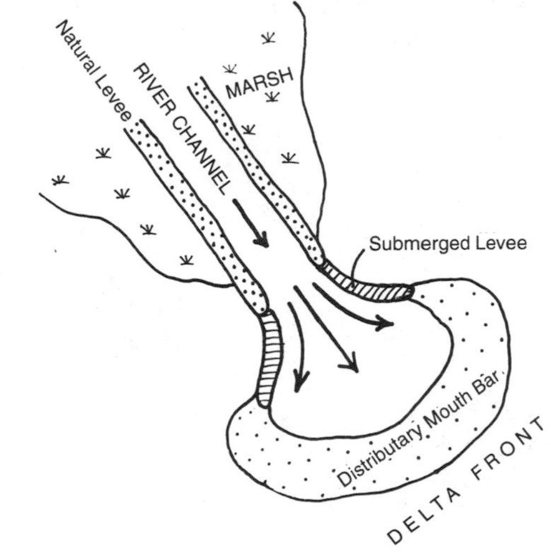

Distributary mouth of a delta.

creasing the area above sea level. Plants eventually transform this new land into marshes.

Second, as the river enters the sea, it moves beyond the confining natural levees, and the flow spreads, dumping an arc of sediment at the river mouth. This bar grows as the stream dumps more sediment and eventually splits the stream, forcing it to flow around both sides of the bar. The river mouth bar splits delta distributaries, after which new river mouth bars grow in each of the new channels, splitting them. And so it goes, widening the delta into its classical fan shape.

By constructing levees, filling bays, and creating river mouth bars, the river and the delta land it creates extend farther out to sea. As the river lengthens seaward, the slope of the channel becomes flatter. And while the delta grows, the river is also building its upriver natural levees higher and

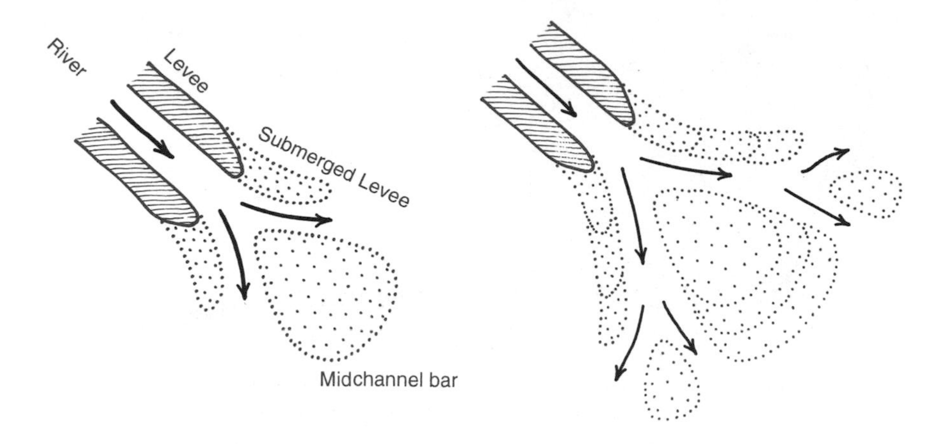

Natural levees at flanks of river grow seaward, extending the delta. Midchannel bar grows, splitting river flow and expanding the delta wider and wider.
—van Heerden and Roberts, 1980

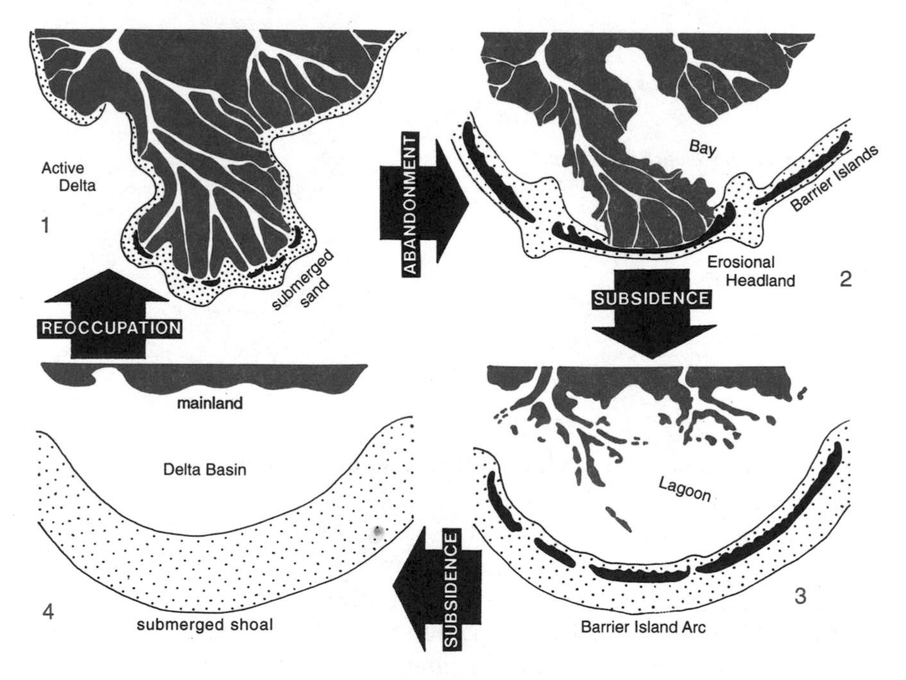

Life cycle of a delta.

1. Delta grows, fed by sediment supplied by active river.
2. River switches, delta is abandoned. Delta subsides and wave erosion attacks delta front. An erosional headland with flanking headlands forms.
3. Subsidence and marine reworking continues. An open lagoon forms inside a barrier island arc. Lacelike delta mainland remains.
4. Only a submerged shoal remains. Delta basin is low, ready to receive a new delta when river switches back again.

—Penland and Boyd, 1985

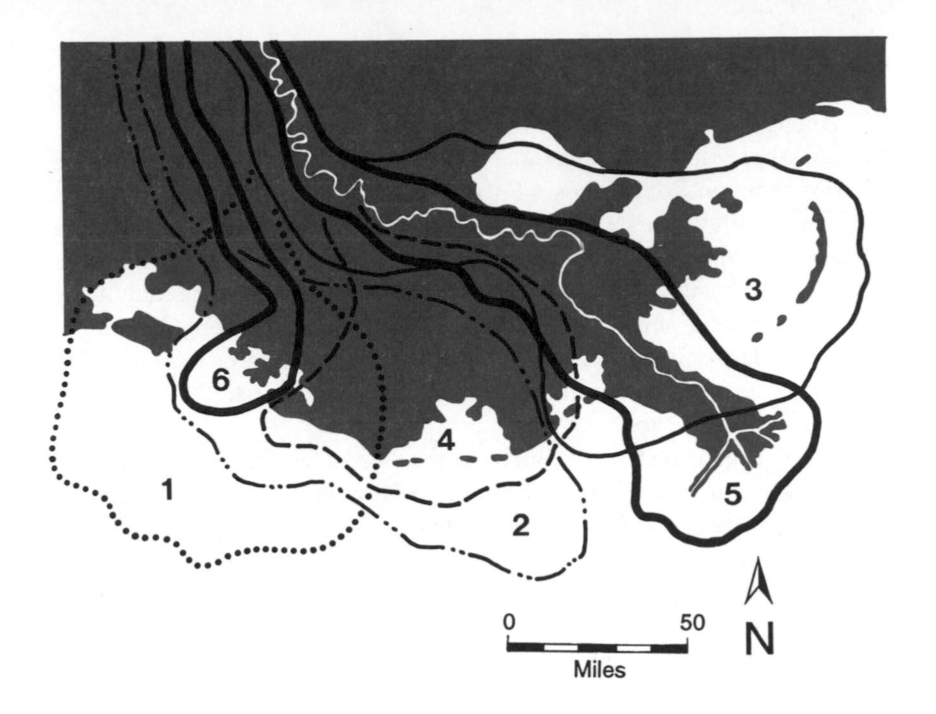

The Mississippi delta has shifted position a number of times during the past 7,500 years. The sequence of deltas, their names (after their main supply river), and ages are: —Coleman, 1988

1 Maringouin (7,500 to 5,000 years ago)
2 Teche (5,500 to 3,800 years ago)
3 St. Bernard (4,000 to 2,000 years ago)
4 Lafourche (2,500 to 800 years ago)
5 Modern Mississippi (Birdfoot) (1,000 years ago to today)
6 Atchafalaya (50 years ago to today)

wider. So as the river extends seaward across the growing delta, it's also confined to a channel within its natural levees. Eventually, the natural levees raise the river above the general level of its floodplain, setting the stage for its diversion into a new channel.

When a large flood finally breaches the natural levee, the water flows downhill onto the floodplain. It can not flow uphill to return to the old channel, so it creates a new channel all the way to the Gulf of Mexico. This shifts the mouth of the river to a new position, where it starts building a new delta while abandoning the old one. That happens again and again.

In the past 7,500 years, the Mississippi River has been furiously building deltas, which began after the vast continental glaciers melted, raising sea level to its present stand. This flooded southern Louisiana in shallow seawater, which the river promptly filled with sediment.

About 7,300 years ago, the main channel of the Mississippi River was located where the present channel of Bayou Maringouin runs. Interstate 10 west of Baton Rouge crosses Bayou Maringouin near Ramah. Between 7,300 and 6,000 years ago, the Mississippi channel built a huge delta southwestward into the Gulf of Mexico, where East and West Cote Blanche Bays and Atchafalaya Bay now are. The main distributary channel of the Maringouin delta came out south of New Iberia, about where Cypremort Point is today. Bayou Cypremort is probably a remnant of the main distributary channel. The Maringouin delta grew for about 2,500 years, until about 5,000 years ago.

Then the main channel shifted to the west side of the valley, into the course of Bayou Teche. The Teche delta partly overlapped the Maringouin delta, but it mainly built seaward a little east of the Maringouin delta. As sedimentation shifted eastward, the Maringouin delta subsided, and open water bays appeared where vast marshes once lay. The Cote Blanche bays and Atchafalaya Bay, Marsh Island, and a few offshore shoals are all that remain of the Maringouin delta. The Teche delta grew for 3,500 years before the river abandoned it and left it to sink.

The main channel of the Mississippi River abruptly shifted east to build the St. Bernard delta into the area where New Orleans now stands. From 4,600 to 700 years ago, the delta grew. After the river abandoned it, waves reworked its front. The Chandeleur Islands, about 70 miles east of New Orleans, and an array of marshes in east St. Bernard Parish are souvenirs of that handiwork.

About 3,500 years ago, the Mississippi River shifted west again, this time running south along the course of Bayou Lafourche. Many remnants of the distributary streams of the Lafourche delta remain as part of the landscape south of Thibodaux. The Lafourche delta grew between 3,500 and 400 years ago, the last of the great deltas that preceded the modern delta. Lake-filled marshes in Terrebonne Parish, Terrebonne Bay, and Timbalier Bay, and the arcuate offshore islands of Isles Dernieres and Timbalier and East Timbalier are relics of the Lafourche delta.

About 1,000 to 800 years ago, the Mississippi River shifted again, into its present Plaquemine course. It then began building the modern birdfoot delta. Chitimacha, Choctaw, and Houma people probably watched the flood that shifted the path of the Mississippi River to its present course. The river built all the land between New Orleans and Venice at the end of the delta in less than 1,000 years.

Some parts of even the modern delta are disappearing beneath sea level because they no longer receive sediment. The lacy pattern of several of the lobes east of the river in the birdfoot delta speak of that subsidence. In fact, the river is trying very hard to abandon this modern delta. Were

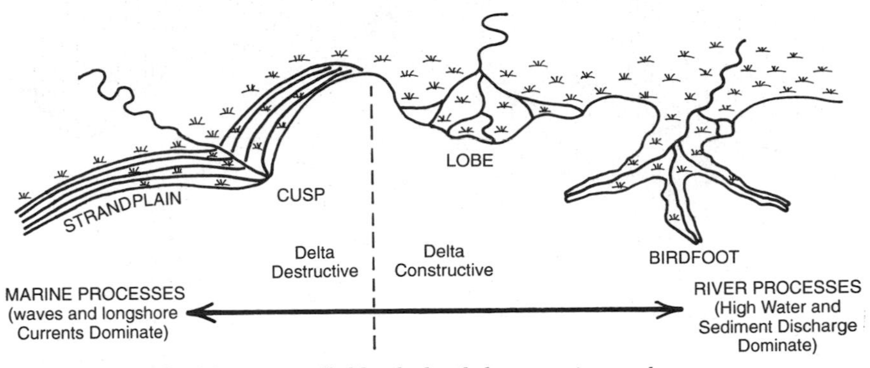

The shape of a delta is controlled by the battle between river and sea. —Scott, 1969

it not for human intervention, it would have switched to the Atchafalaya channel in 1973.

The Atchafalaya River, the main distributary channel of the Mississippi River, heads straight south to the Gulf of Mexico from the Old River near Simmesport. This path is 190 miles shorter and 12 feet lower than the present looping course of the Mississippi River around New Orleans and down the birdfoot delta. It would be natural for the Mississippi River to switch to the Atchafalaya channel now. Since 1850, the Atchafalaya River has been filling the lakes in its lower stretches with sediment. After the great flood of 1973, the first hints of a new delta rose above sea level in Atchafalaya Bay. Were it not for the Old River control structure, the Atchafalaya River would now be the main channel.

Mississippi Fan

During the ice ages, when sea level was at times as much as 450 feet lower than it is now, large areas of the seaward part of the delta were above sea level. The Mississippi River flowed across the exposed delta, and dumped its load of mud down the submarine slope to the deep ocean floor, where it piled up to build the immense Mississippi fan, an enormous cone of

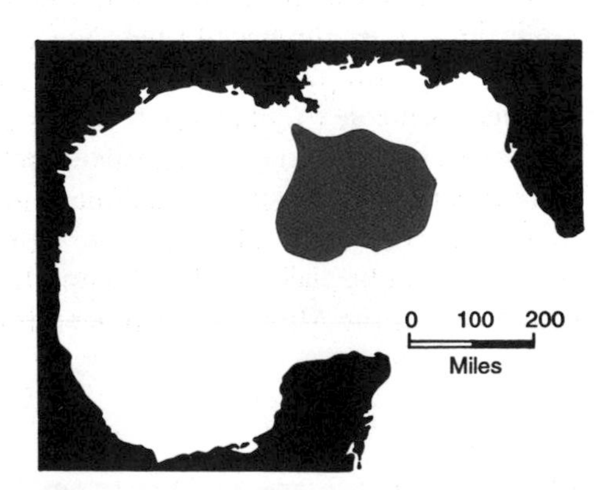

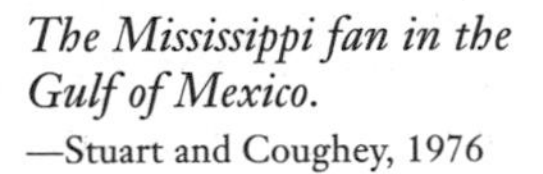

The Mississippi fan in the Gulf of Mexico.
—Stuart and Coughey, 1976

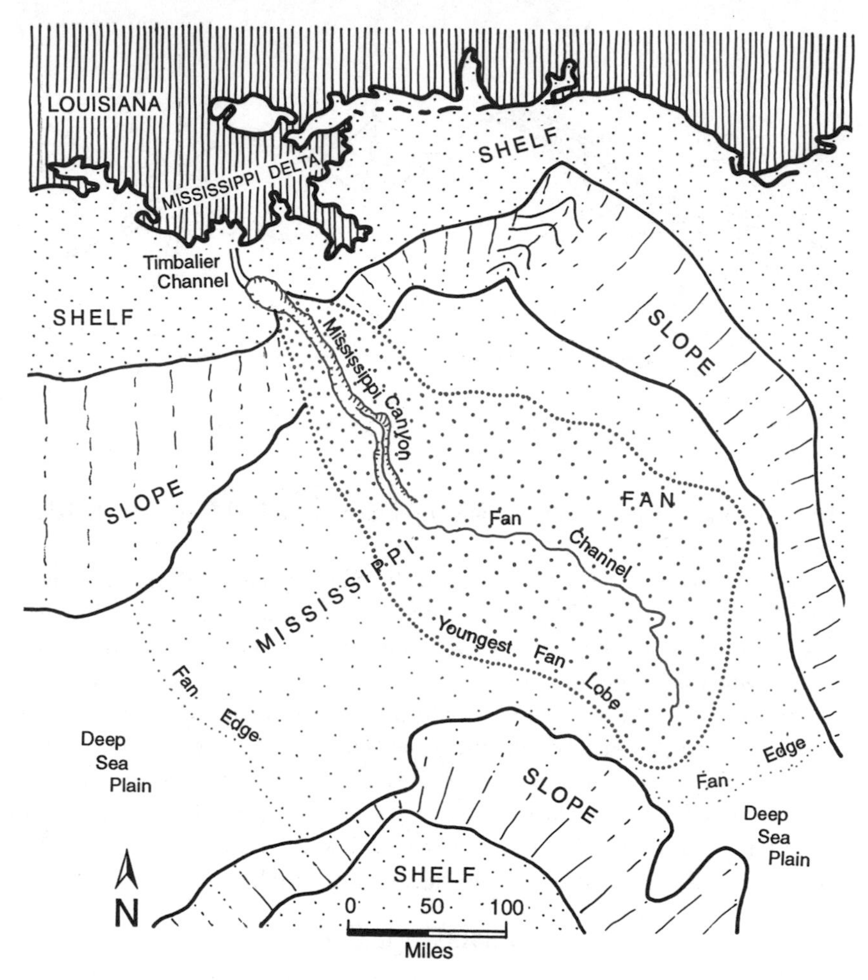

The Mississippi fan. —Boyd et al., 1989

muddy sediment that nearly covers the floor of the Gulf of Mexico. The fan is three times the size of Louisiana, and 8,000 feet thick near its upper end—a submarine delta, built mostly of river sediments in Pleistocene time.

Sea level rose to submerge the seaward part of the delta when the last of the great glaciers melted, approximately 12,000 years ago. The Mississippi River then deposited its sediment load where it met the shallow sea, high up on the delta. This was the beginning, about 7,500 years ago, of the modern delta plain and Mississippi that you now see. Even though you cannot see it, the Mississippi fan is an important part of the story of the Louisiana landscape.

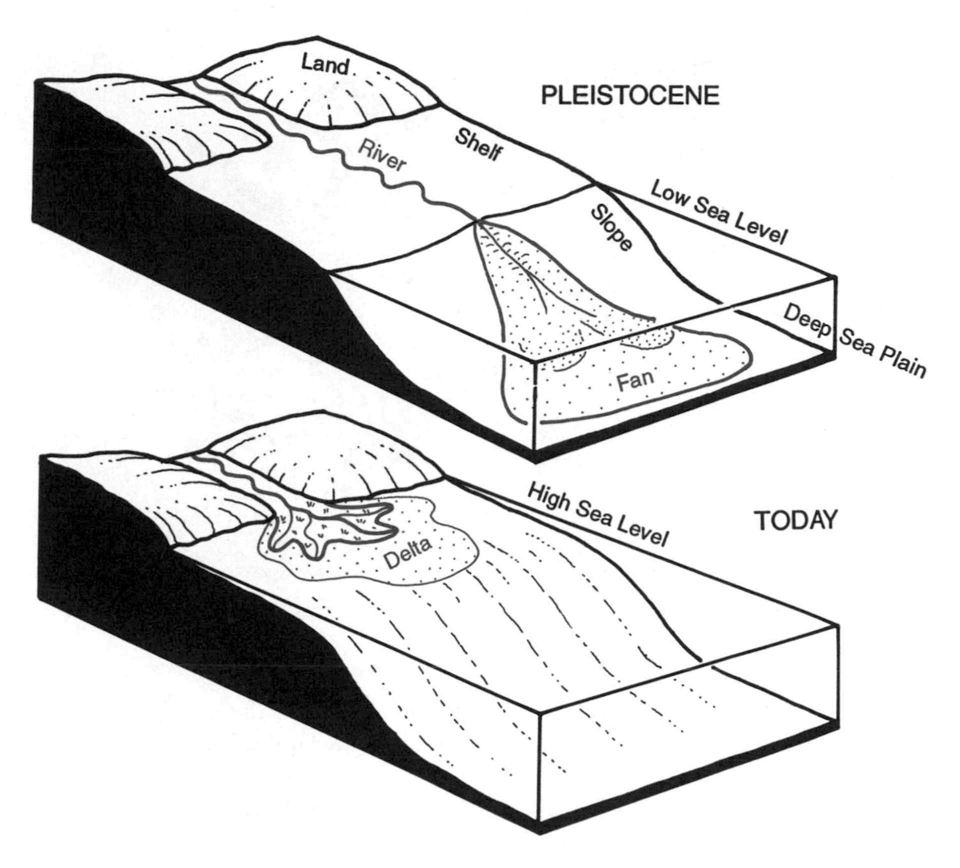

During low sea level periods of the ice age (Pleistocene), the Mississippi River delivered its sediment load to the deep sea of the Gulf of Mexico, forming the Mississippi fan. When ice melted and sea level rose (about 7,500 years ago), the Mississippi began to build a shallow-water delta near land.

How the Mississippi River Grew

The Mississippi River and its tributaries collect water and sediment from 32 states and 2 Canadian provinces. Water from the west slope of the Appalachians joins runoff from the Rocky Mountains. It gathers the rainfall of a third of the country, from more than a million square miles of drainage basin to carry some 375 billion gallons a day through Louisiana.

The Mississippi River could be 500 million years old, if some maps are correct that show a major river flowing from Canada to a delta at the southern edge of the Ordovician version of North America. The river's origin 200 million years ago could have begun when North and South

The Mississippi drainage basin covers over 1 million square miles, collecting water from 32 states. —Schumm and Brackenridge, 1987

America drifted apart, opening the Gulf of Mexico in the widening gap between them. If so, it would have followed the lowland rift, a ragged tear in the crust that sliced northward as the Gulf of Mexico opened. For the next 198 million years, the Mississippi River contributed enormous amounts of sediment to the edge of the continent, building the edge of North America nearly 250 miles into the Gulf of Mexico. It added all of Louisiana during the last 65 million years.

The Mississippi River of 2 million years ago was much smaller than the stream we know. Western North America was extremely dry between about 15 and 2 million years ago, when the first of the great ice ages began. Too little rain fell in the western headwaters of the Mississippi River to maintain permanently flowing streams. Vast quantities of sediment were left in the western drainage basins and spread across the high plains. The climate was wetter at the beginning of Pleistocene time; permanent streams flowed into the Mississippi from the west, and they eroded the sediment that had been left behind in the dry years, and began hauling it to Louisiana.

Before the ice ages, the Mississippi River drained only the southern third of the Midwest. Much of the water from farther north drained east to the Atlantic Ocean and north to Hudson Bay. Then the great glaciers that covered most of Canada and the northern tier of the United States several times during Pleistocene time blocked drainage to the east and north. When the ice melted, the streams continued to flow south, adding

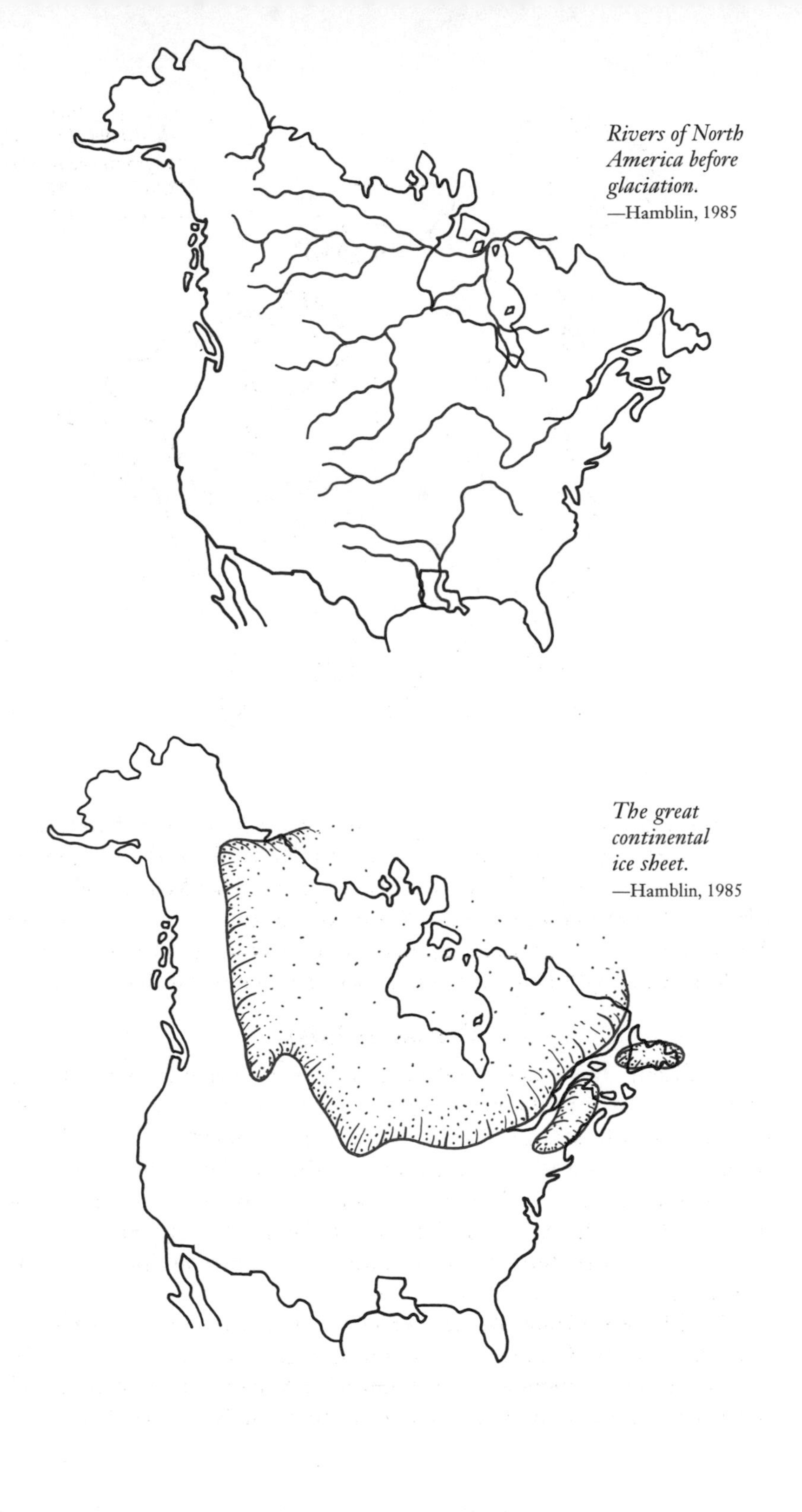

Rivers of North America before glaciation.
—Hamblin, 1985

The great continental ice sheet.
—Hamblin, 1985

After glaciation the Mississippi River became the preeminent North American river.
—Hamblin, 1985

an enormous area to the drainage basin of the Mississippi River. They carried with them vast loads of sediment released from the melting ice. The river deposited some of those sediments to make Macon Ridge near Monroe. Other deposits exist in the gravels in the channel bottoms the Mississippi River cut into older rocks. They also exist in the great bands of Pleistocene sediments that spread across much of the state.

Wetlands Loss

Louisiana hosts nearly 40 percent of the wetlands of the United States, and they are disappearing at a rapid rate. About 35 square miles of wetlands in Louisiana slip away each year, an important loss of habitat to innumerable species of birds, animals, fish, and plants. Public attention has focused on wetlands loss in Louisiana in recent years, and blame flies in many directions. In truth, the phenomenon is largely due to natural geologic processes, though human intervention certainly plays a significant role.

Vanishing wetlands are neither new nor unusual in Louisiana. The marshland plains of southern Louisiana developed during more than 7,000 years of delta construction. Rivers build deltas first here, then there, switching from one to another as a major channel builds a delta then diverts

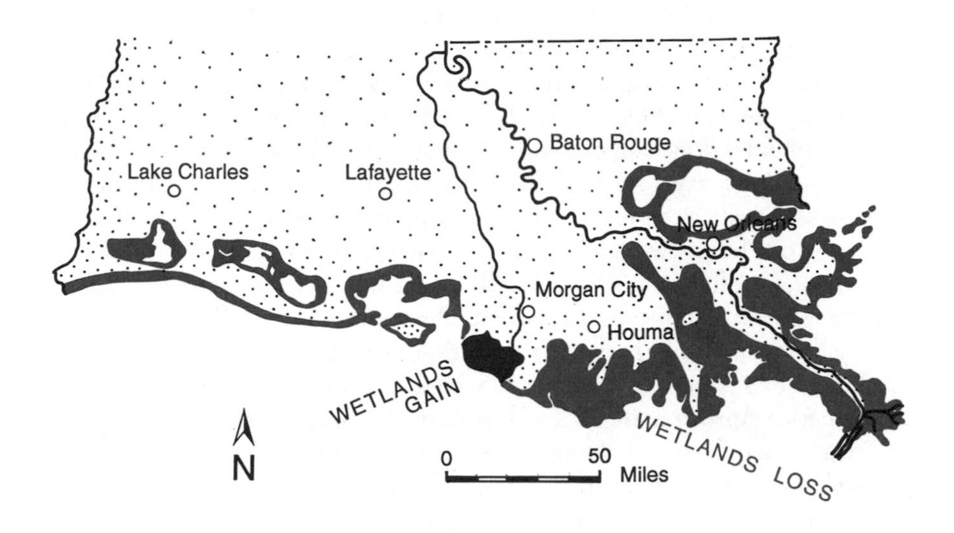

Present areas of wetlands loss and gain on the Louisiana coastal plain. —Steward and Berry, 1990a

through a crevasse to build another. That leaves the old delta abandoned, without a continuing supply of sediment. Its wetlands slowly subside beneath sea level as waves rework its front into a line of barrier islands. Vast areas of marshy wetlands have been built, abandoned, and lost to subsidence during the past few thousand years. They sank beneath the sea long before humans ever dredged a channel, built a control levee, or cut through a marsh to reach an offshore oil platform.

Whether deltas and their wetlands survive or sink beneath the sea hinges on the fine balance between sediment and subsidence. Deltas tend to sink as their soft and watery mud compacts beneath the weight of more mud laid on it. If no new sediment is added to the top of the pile, the top of the delta will sink beneath the waves. If the supply of sediment delivered to the delta exceeds the sinking rate, the delta will continue to grow and wetlands will expand. Channel shifts and delta switches cause wetlands to grow around the mouth of the new channel while wetlands disappear on the abandoned part of the delta, now deprived of sediment. This geologic process remains the chief reason for disappearing wetlands in southern Louisiana.

Nevertheless, tinkering human hands have certainly accelerated wetlands loss during the past century. Their principal contribution has been construction of continuous levees along the Mississippi River from Baton Rouge to Venice. The levees do contain floods, but they also deprive the marshes of the sediment they need to stay above sea level. Furthermore, the levees force the river to dump all its sediment at the extreme end of the delta, where waves and currents cannot rework it into sandy coastlines as they did before the levees were built. Much of the sediment slides down the continental slope and onto the Mississippi fan on the deep ocean floor.

Several schemes are afoot to divert some of the sediment through pipelines, gated levee breaks, and open gaps in the lower delta to replenish selected patches of wetlands, but the cost will be high. Moreover, only fine-grained sediment will be diverted, not the heavy, delta-building bedload sliding along the bottom of the main channel.

The second major human contribution to wetland destruction through sediment loss is outside Louisiana. Scores of dams were built during the past century upriver in the Mississippi drainage and in its tributaries. They were designed to control floods, generate electricity, store fresh water, and provide recreation. They also trap so much sediment that what reaches Louisiana is only about half that of a century ago.

Other human activities contribute in smaller ways to marsh destruction. Many canals that have been dredged across fresh marshes, mainly for navigation, pipeline construction, and barge access to petroleum drilling platforms, accelerate saltwater intrusion. Wells that produce fresh water, oil, and gas locally contribute to subsidence. Additional natural processes also help destroy some wetlands. Hurricanes damage barrier islands, which protect marshes from wave erosion. The current extremely slow rise in sea level helps inundate fresh marshes with salt water.

All these factors contribute to wetland loss. One bright exception to the general wetland reduction is the Atchafalaya delta, which is adding land at an astonishing rate in Atchafalaya Bay.

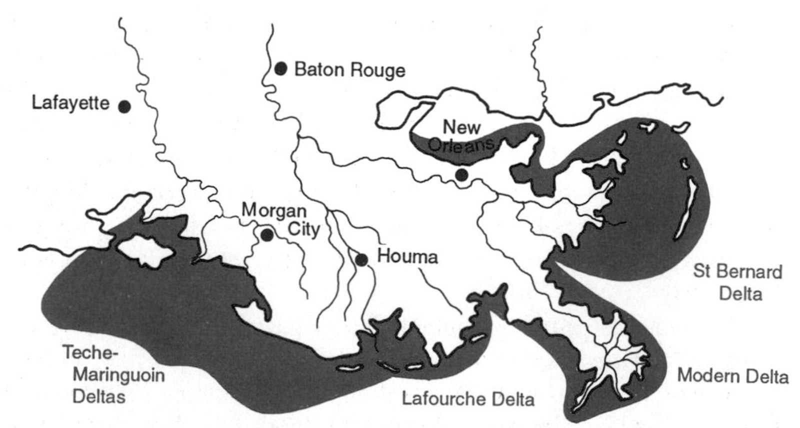

Areas of land loss during the past few thousand years from delta shifting and subsequent destruction. —Steward and Berry, 1990b

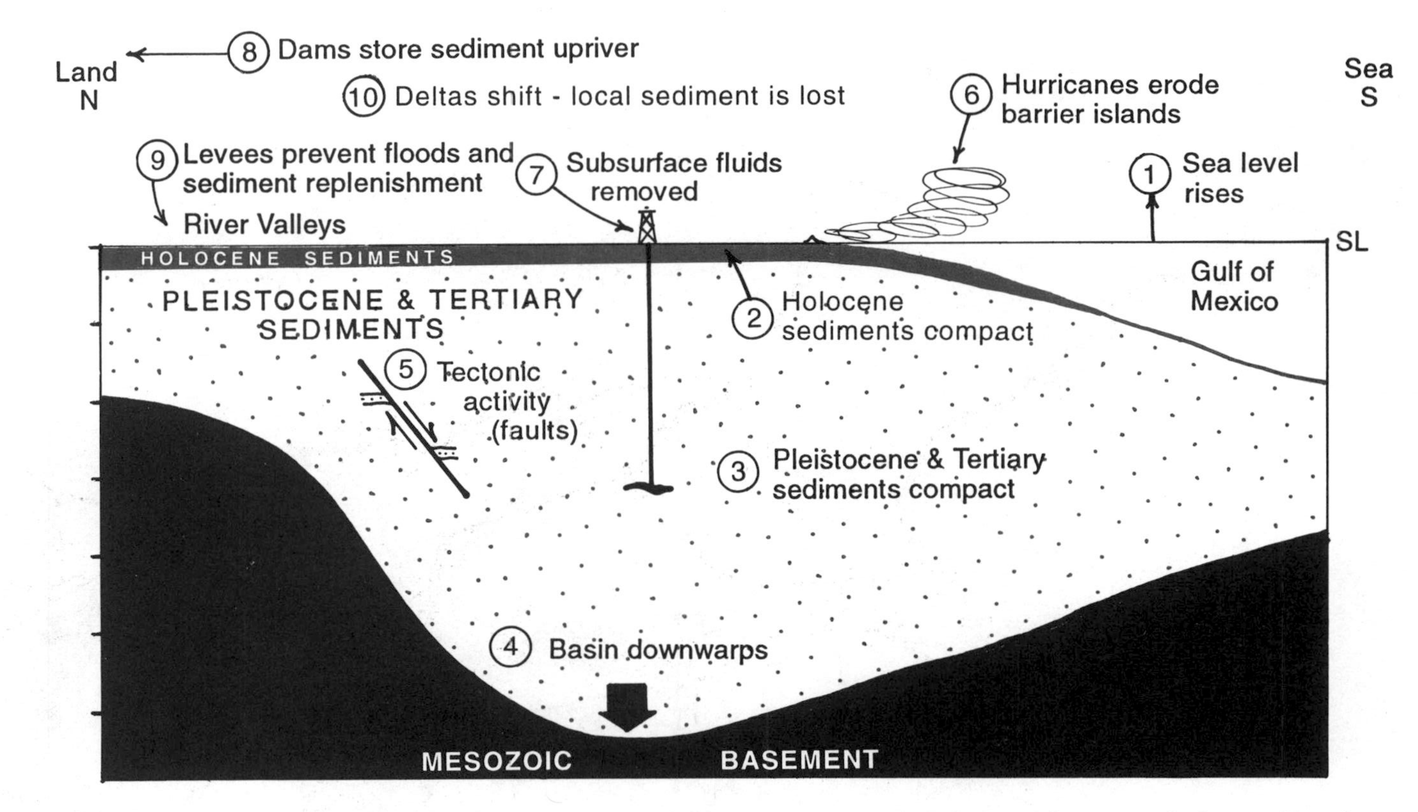

Multiple factors, both natural and man-made, contribute to wetland loss in Louisiana. Principal natural factors are subsidence and loss of sediment from delta shifts. Chief human factors are levees and dams. —Davis, 1990

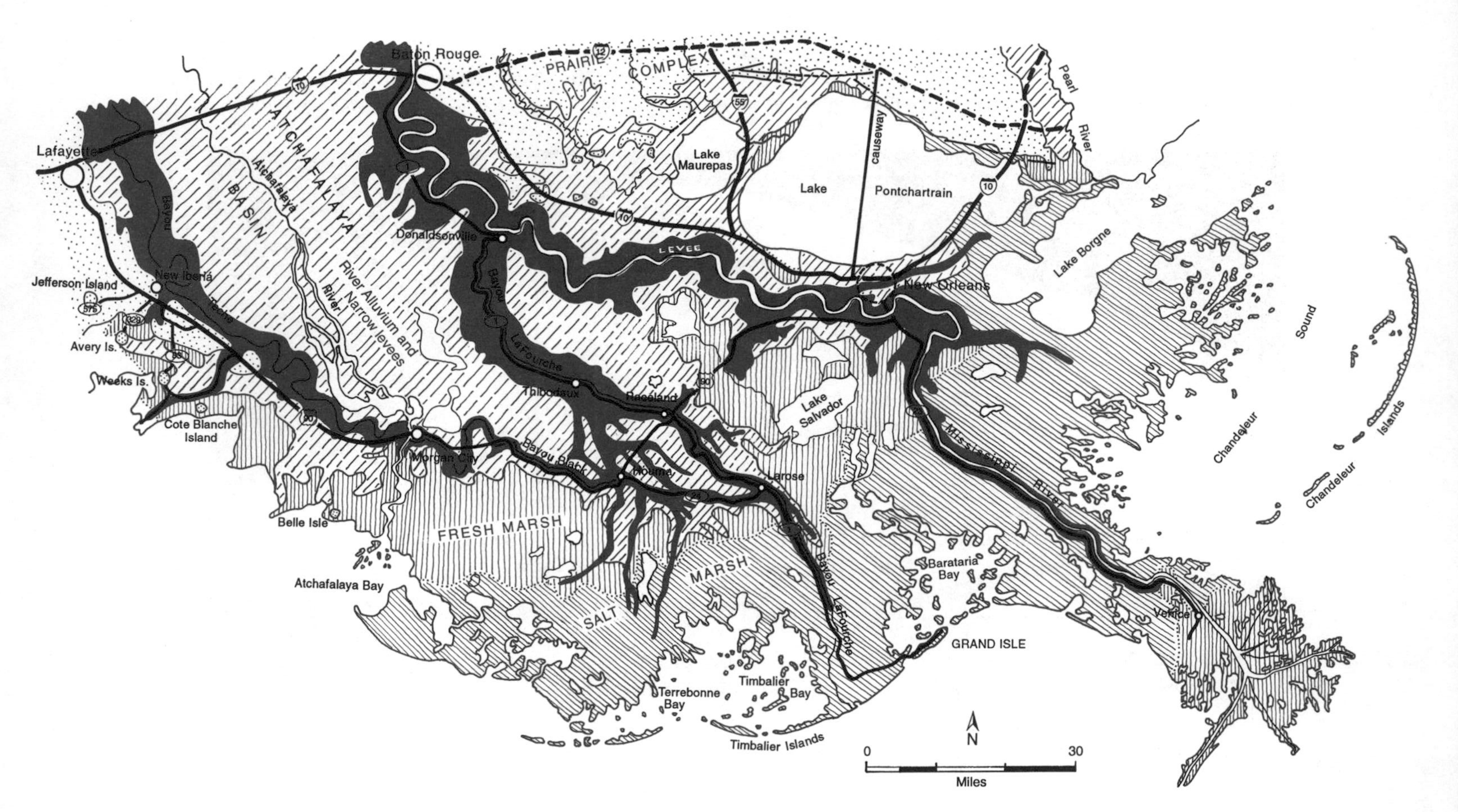

Geologic map of southeastern Louisiana.

Southeastern Louisiana

Uneasy Rivers, Marshes, and Deltas

The large population of southeastern Louisiana, living on one of the most dynamic landscapes in the world, accentuates human interaction with the forces of nature as perhaps no other place can, not even California with its earthquakes. Southeastern Louisiana has its moments of drama, too, especially when occasional floods and hurricanes roar down on delta dwellers. But the changes wrought on the delta by the relentless deposits of mud and silt, the continual wave erosion of marshes, and the inexorable subsidence create a drama on a scale unmatched in virtually any other natural environment.

You can drive the length of the modern Mississippi delta on Louisiana 23 south of New Orleans. Explore the history of Lake Pontchartrain and its environmental crisis as you cross the Pontchartrain causeway out of

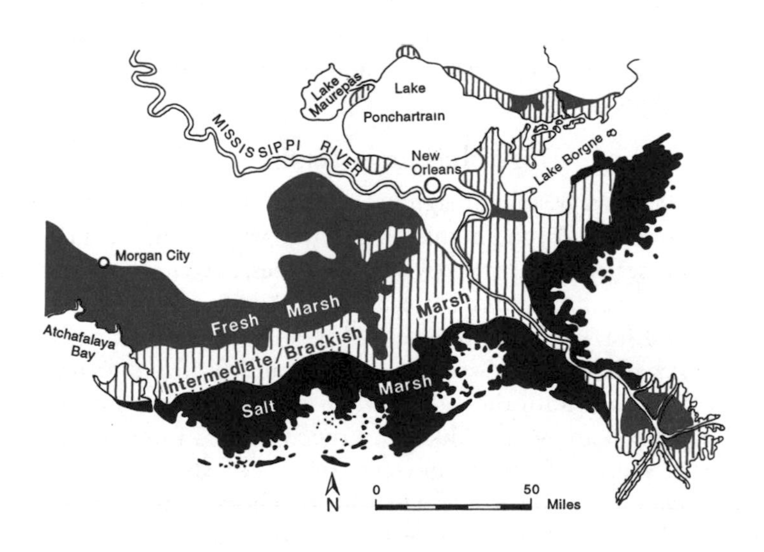

The marshes of the delta plain. —Gagliano et al., 1981

The Creole Queen.

New Orleans. Discover how buildings are designed in New Orleans to withstand subsidence, and where all the water goes when it rains in the Crescent City, which is 12 feet below sea level in places.

At several of the famous Five Islands near New Iberia, salt pinnacles have penetrated nearly to the marshy surface, pushing island hills nearly 100 feet above the marsh. Notice the remnants of the old Lafourche delta near Houma, where once the Mississippi River ran. At the Atchafalaya River basin, Cajun folk, crawfish delicacies, and a phenomenal land and river come together. You can see the basin at treetop level from the elevated causeway of Interstate 10. Think about changing shorelines, lost barrier islands, and the dynamic coastline as you drive the length of Louisiana's premier sandy barrier island, Grand Isle, on the Gulf Coast south of Leeville.

Atchafalaya Basin

The Atchafalaya basin is a magnificent wetland wilderness, where much of the story of wetlands, and human interaction with them, is being played out on an epic scale. The cypress swamps, marshes, bayous, lakes, and rivers of the Atchafalaya basin stretch for 35 miles between Baton Rouge and Lafayette, and for 125 miles between Morgan City and Simmesport.

The Atchafalaya basin was the hunting and fishing domain of the Chitimacha Indians until the 1700s, when French explorers drove them deeper into the swampy bottomlands. French settlers from Nova Scotia, the Acadians, settled on prairies and levee ridges near the Atchafalaya basin later in the century and developed a new life that became indigenous Cajun. But it is the landscape we are interested in here, and the story of the land

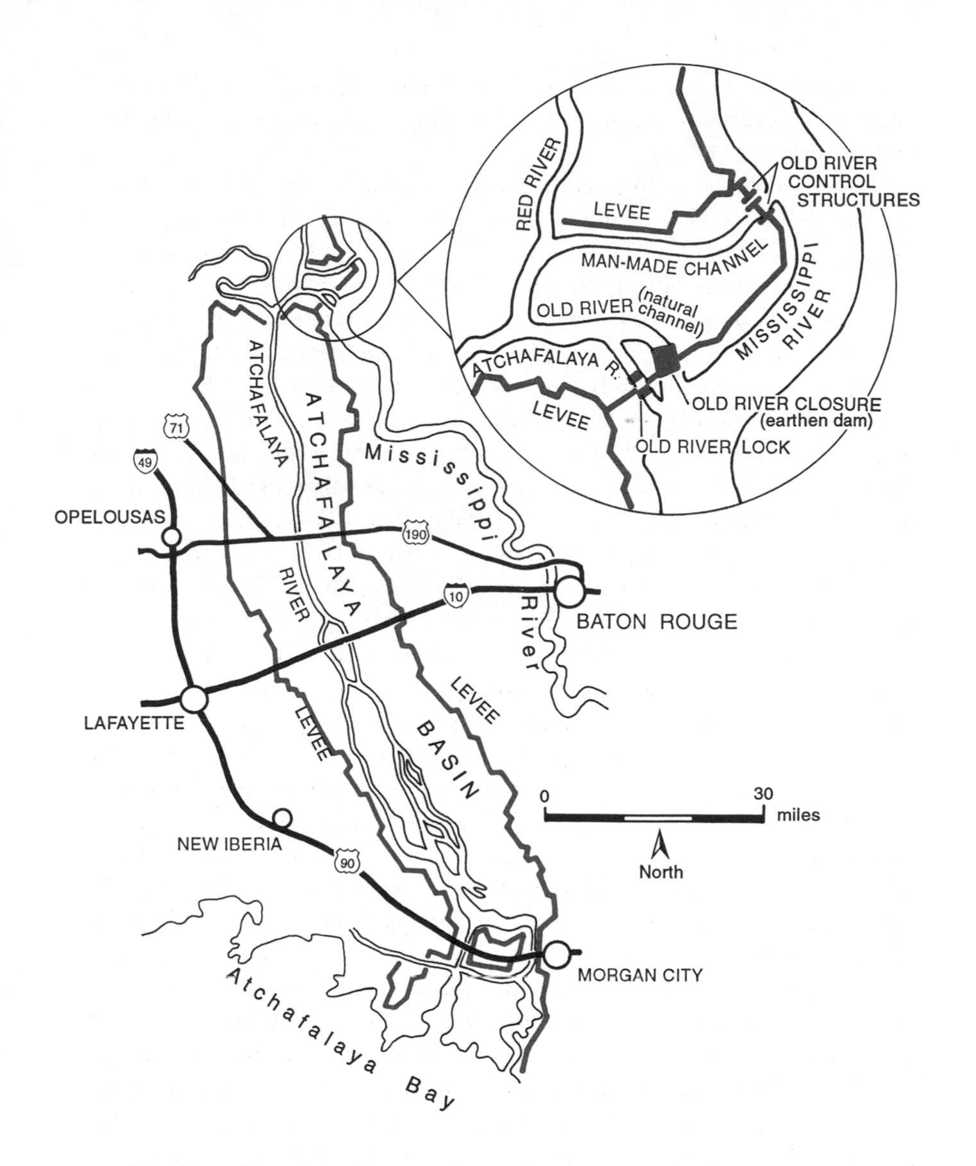

Atchafalaya River basin levees and control structures. —U. S. Army Corps Engrs., 1978; Louisiana Hydroelectric Co.

is fascinating. Some geologists call the Atchafalaya basin the "phenomenon of the twentieth century," for this land has changed remarkably fast, and continues to change.

The Atchafalaya River is the main distributary channel of the Mississippi River, which is trying to switch into the Atchafalaya's steeper and shorter path to the Gulf of Mexico. The Mississippi River has switched course several times during the past 7,500 years, so taking the Atchafalaya route would continue this natural pattern, which would save the river about 190 miles. In the 1950s, geologists predicted the Mississippi River would naturally switch to the Atchafalaya River sometime in the 1970s, unless something was done to prevent it from happening. Something was done, of course: The Old River control structure was built northwest of Baton Rouge in the 1960s to divert 30 percent of the Mississippi River, along with all of the Red River, into the Atchafalaya River. Too much commerce, navigation, and freshwater supplies were at stake to allow the entire lower Mississippi River to divert into the Atchafalaya. The loss of fresh water alone would be catastrophic to New Orleans and the river industries. However, the loss of silt would likely turn the Mississippi channel into a deep estuary for many years and would save the Corps of Engineers endless dredging to keep the shipping channel open to Baton Rouge.

In the 1500s, the Mississippi River captured the lower Red River, and established the Atchafalaya River as a distributary. The capture was probably done by a small nameless tributary that headed in the Moncla Gap area and flowed into the Mississippi River at Turnbull's Bend. The Red River followed its own wide floodplain valley and never flowed directly into the Atchafalaya River before the cutoff of Turnbull's Bend. The Atchafalaya probably developed as a stream draining the alluvial basin between active Mississippi River meander belts, then eroded headward until it captured a crevasse channel that radiated from the Mississippi River at Turnbull's Bend. This turned the Atchafalaya River into a distributary of the Mississippi.

Then a huge log jam developed at the head of the Atchafalaya River, which kept it from taking too much water from the Mississippi River. That maintained the Atchafalaya River as a stream, rather than a substantial river. Much of the lower Atchafalaya basin was then a depression that held large lakes. This was the natural setting of the Atchafalaya basin until the 1800s, when energetic people started the long process of messing with the river.

In 1831, Captain Shreve dug his shortcut across the neck of Turnbull's Bend on the river to shorten the route for steamboats. It was considered a marvelous engineering feat, but no one could then foresee its consequences: The cutoff sent more of the muddy waters of the Red River down the Atchafalaya River, greatly increasing sediment supply to the Atchafalaya basin, while shortchanging the Mississippi River. Though the

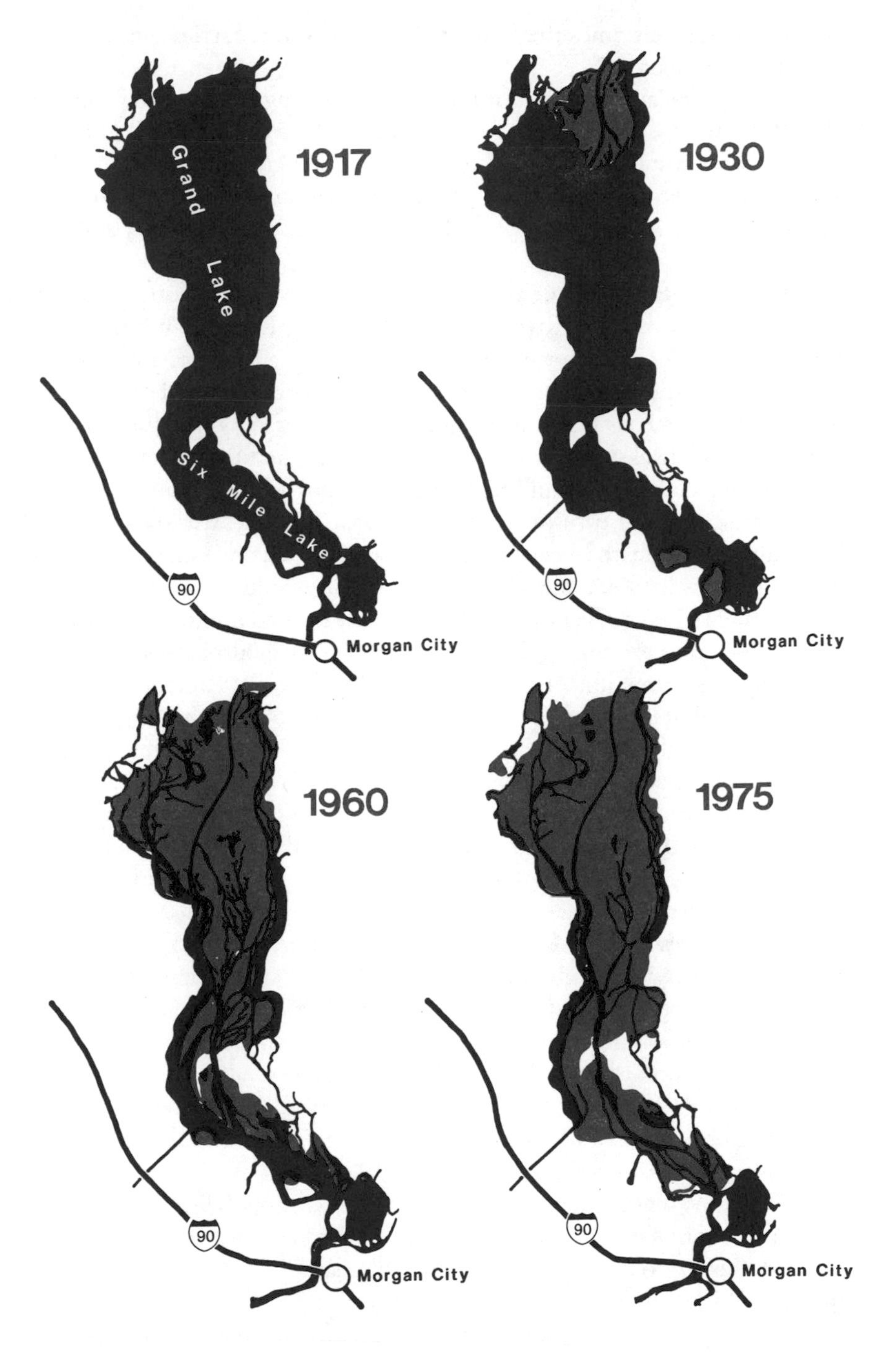

The sediment-laden Atchafalaya River has been building toward the Gulf of Mexico by filling in Grand and Six Mile Lakes in the river's watery reaches north of Morgan City. —Roberts et al., 1980

Mississippi has ten times the flow of the Red River, it carries one-quarter as much sediment.

A few years later, Shreve got the job of removing the log jam from the Red River to free the channel for navigation. This caused the Red River to deliver even more water and sediment to the Atchafalaya River. The state of Louisiana arranged a contract in 1839 to remove the log jam from the head of the Atchafalaya River. By 1855, the logs were gone, and the stage was set.

The Atchafalaya River promptly took more water from both the Mississippi and the Red Rivers, especially during floods. The increased flow enlarged it into a formidable river carrying a gigantic load of red sediment from the Red River. In the next 100 years, the enlarged Atchafalaya River transformed its basin: it straightened its course by building natural levees, it filled backswamps, and it built deltas at the heads of lakes in the lower basin. Grand Lake and Six Mile Lake north of Morgan City were open water near the turn of the century. By 1930, the Atchafalaya River had built a delta at the head of Grand Lake. Thirty years later, most of Grand Lake was filled, and Six Mile Lake also would soon fill.

By the 1950s, geologists and river engineers had studied the history of Louisiana's rivers and understood enough of river processes to recognize what was happening. They saw that the Atchafalaya River was a major river, a distributary of the Mississippi. They understood that the Mississippi River would soon follow the Atchafalaya River in a shorter path to the sea. It had, after all, switched its course many times in the past. But that natural turn of events would switch the mouth of the river from below New Orleans to Morgan City. It would leave all the infrastructure of navigation improvements and wharfage from Baton Rouge south isolated on a salty estuary.

In the 1960s the Old River control structure was built to regulate the flow of Mississippi water into the Atchafalaya River. It would also be possible to open the control gates during especially high floods to let water spread out through the swamps, bayous, and lakes of the Atchafalaya basin to spare downriver communities the worst of the flood. Levees built along the length of the Atchafalaya basin controlled the lateral spread of floodwater, but the basin retained much of its natural character, and even today development is minimal.

One of the biggest floods came down the Mississippi River in 1973. The Old River control gates were opened wide, and water and sediment blasted through them into the Atchafalaya River, the raging waters nearly destroying the Old River structure. The gates shook violently, but they held.

The floodwaters poured down the Atchafalaya River to Atchafalaya Bay, where they dumped their sediment. As the water receded, small

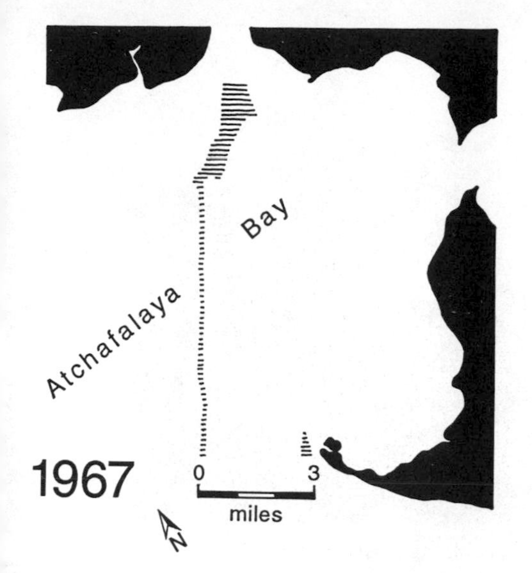

In 1973 the Atchafalaya River had succeeded: its first abovewater delta segments appeared in Atchafalaya Bay, following the big flood of that year. The delta continues to grow.
—Roberts et al., 1980

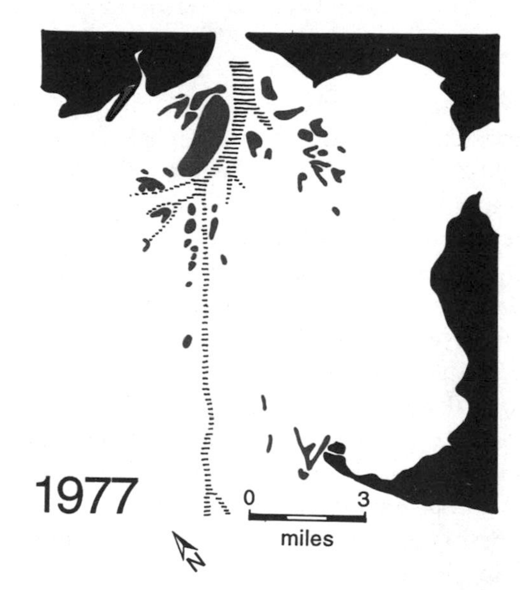

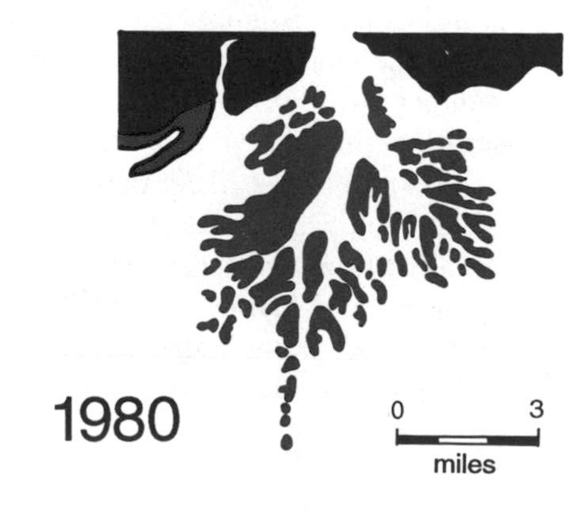

Looking out over the Mississippi River from Jackson Square in the French Quarter, you can see the river is higher than the street level!

islands of a new delta appeared in the bay. Though mud had been accumulating underwater at the mouth of the Atchafalaya River for years, the 1973 flood finally delivered enough to begin a delta. While wetlands were sinking and disappearing elsewhere in Louisiana, the Atchafalaya River was building new land. It had filled its basin and was delivering sediment to the sea. Geologists estimate Atchafalaya Bay will fill in the next 50 years, making marshlands where open water now stands.

What would have happened if Captain Shreve and his successors had not tampered with the Mississippi River? Would it have naturally switched to the Atchafalaya River course? Would the Atchafalaya River have filled its basin and created a new delta? The answer to both questions is, probably. Both basin filling and delta building are natural processes and would have occurred without human interference.

Much discussion and controversy centers on management of the Atchafalaya basin. Some people advocate letting nature take its course, allowing the Mississippi River to switch into the Atchafalaya channel. Others decry the changes that would happen to the Atchafalaya basin, causing sediment to fill open water habitats. The issues are complicated and not simply resolved.

It seems most likely that the present course of the Mississippi River will be maintained as the main channel—too much infrastructure is involved to allow the switch. Efforts are made to protect the wilderness character and habitats of the Atchafalaya basin. Water is now being diverted into many backswamps and small lakes to prevent them from drying up. It does seem a bit ironic that we question the processes that now fill the Atchafalaya basin with sediment and create a growing marshy delta as concern mounts for the loss of marshes and wetlands elsewhere in Louisiana.

New Orleans

The Crescent City is a seaport city below sea level. During spring floods, when the levees are filled, giant tankers ride by on the other side of the levee with their waterlines dozens of feet over your head. During the Civil War, Admiral Farragut's warships, riding high on the spring floods, pointed their guns down on the city and forced its surrender without firing a shot.

From its founding in 1718, when Bienville established his French colony on the natural levee bordering the Mississippi River, New Orleans has battled the river, swamps, flooding, and sinking to stay above water. It would

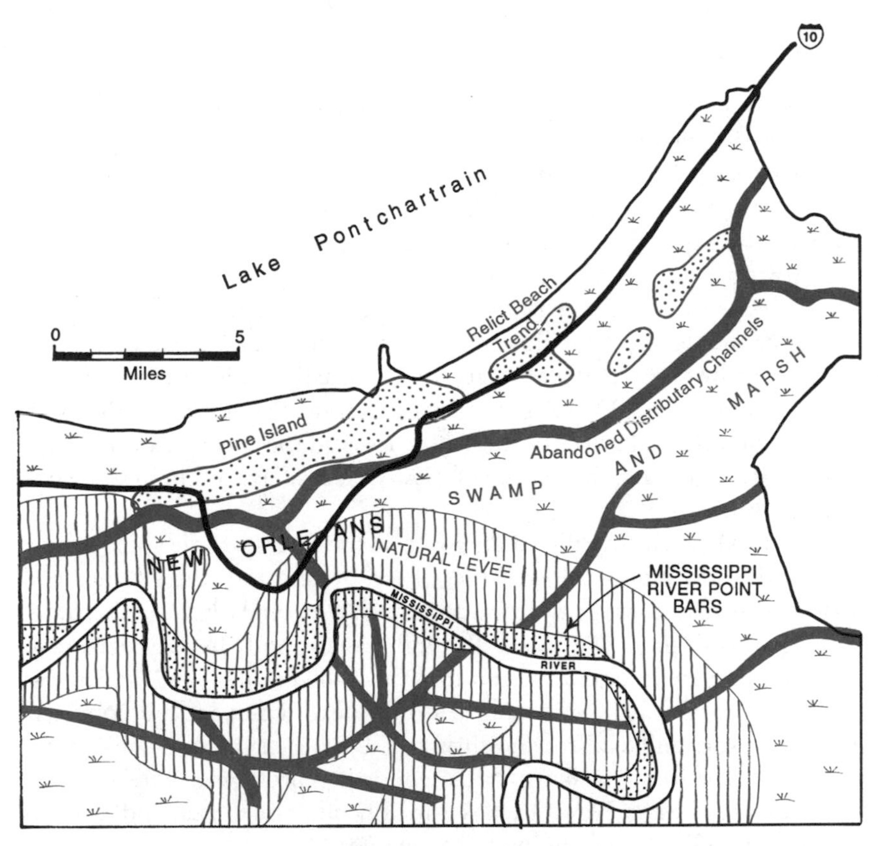

Geologic map of greater New Orleans. Modern natural levees, standing nearly 15 feet above sea level, were obvious sites for early settlement. Modern swamp and marsh were later drained and developed; drying peat can cause local subsidence. Deposits just beneath the surface, such as point bars, abandoned distributary channels, and relict beach sands provide firmer footings for construction. —Snowden et al., 1980

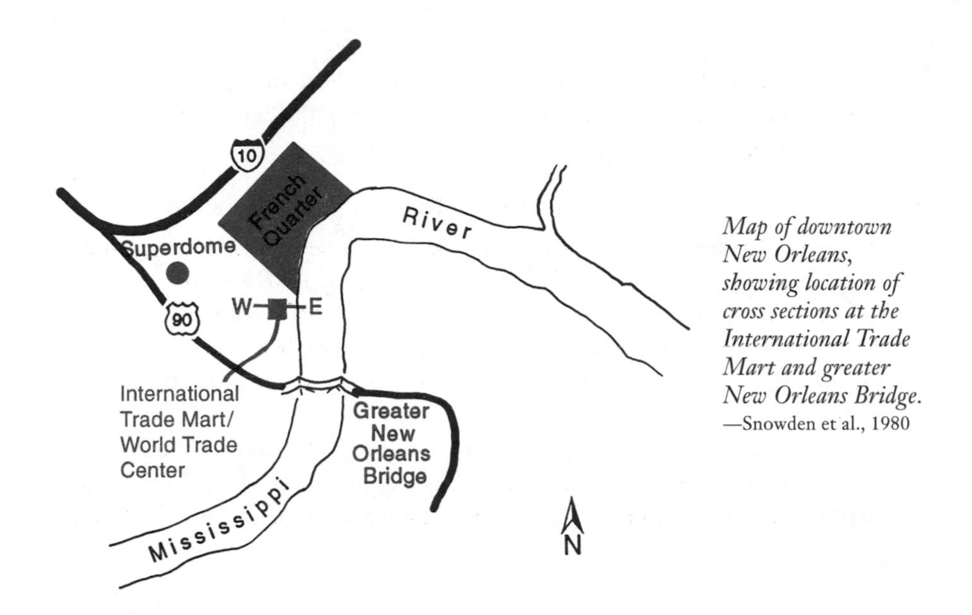

Map of downtown New Orleans, showing location of cross sections at the International Trade Mart and greater New Orleans Bridge.
—Snowden et al., 1980

be hard to find a more difficult place to build a major city. The soft mud and peat that lie beneath most of the city are not what engineers dream of when they plan a heavy office tower or freeway overpass. How does this metropolitan giant cope? How do big buildings stand without toppling into a muddy quagmire? And why is the city not under water, considering that parts are 12 feet beneath sea level?

By the time the French Quarter was laid out 15 feet above sea level on the natural levee, Bienville's engineers had built an artificial levee 3 feet high to further protect the new city. By 1735, artificial levees extended 30 miles upstream and 12 miles downstream from New Orleans. By 1812, continuous artificial levees bordered the river all the way to Baton Rouge. For nearly 300 years, the people of New Orleans have attempted to control the water.

By 1900, when virtually all the space on the levees was occupied, newly designed water pumps allowed the city to expand into the swamps and marshes north of town, between the natural levees and Lake Pontchartrain. The pumps drained the water through canals, and the city expanded northward. New levees were built to protect the drained land, which was covered with as much as 15 feet of sloppy peat. The peatland compacts under pressure and oxidizes and shrinks if dried, causing local subsidence, cracked foundations, sinking yards, and other problems. The Kenner-Metairie area west of New Orleans is mainly on drained peatland, and has subsided rapidly during the past few decades.

More favorable natural foundations are used for large construction projects such as freeways and buildings. For example, Interstate 10 east of

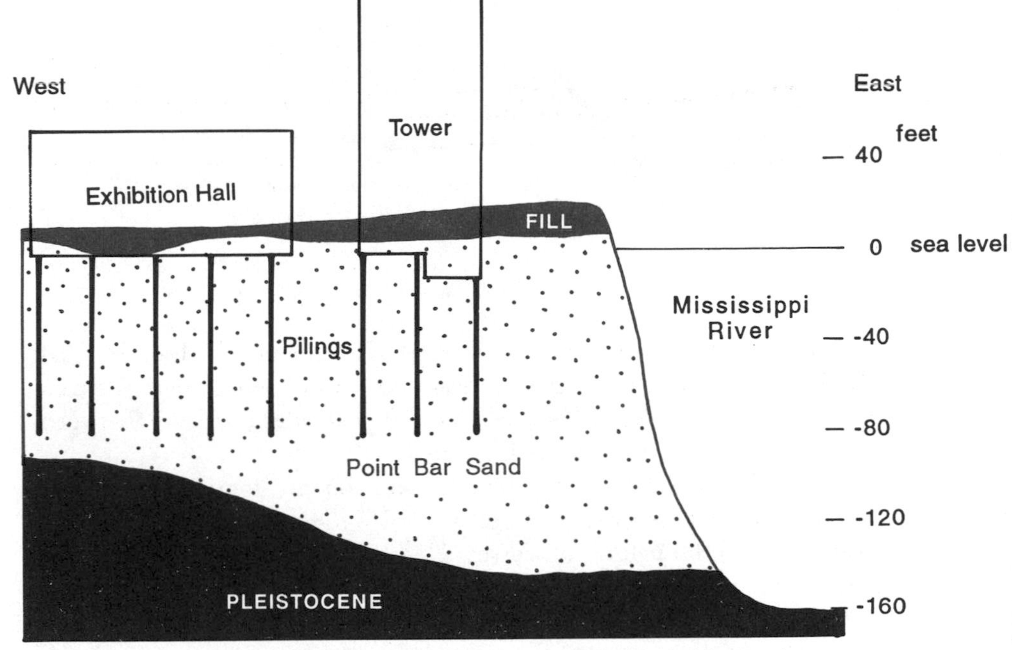

Cross section of the geology beneath the International Trade Mart at the foot of Canal Street, showing how the buildings are supported by pilings driven into firm sand. —Snowden et al., 1980

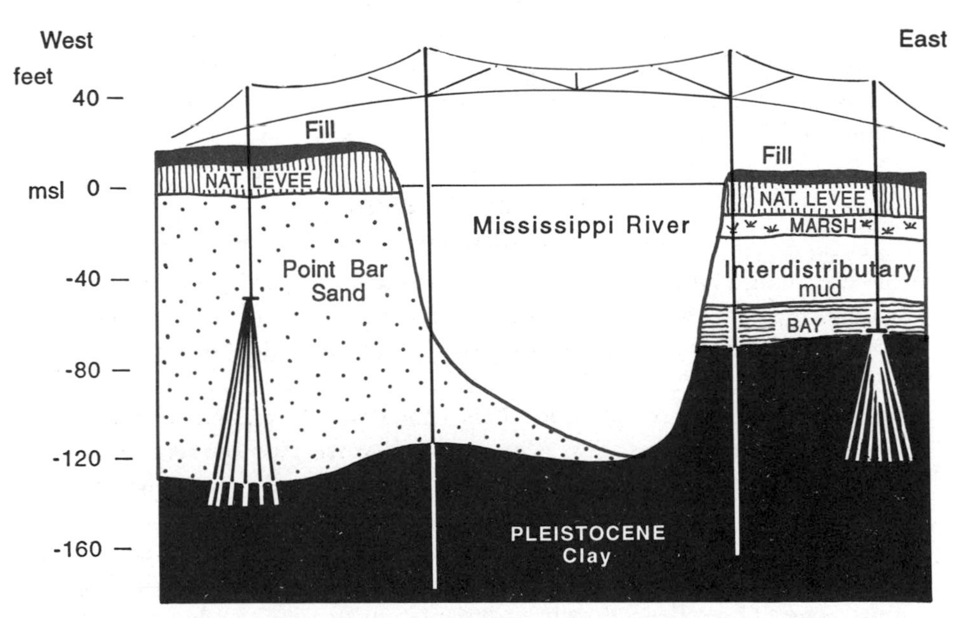

Cross section of the geology beneath the greater New Orleans Mississippi River Bridge (US 90), showing pilings seated in stiff Pleistocene clays (considered bedrock in New Orleans). —Snowden et al., 1980

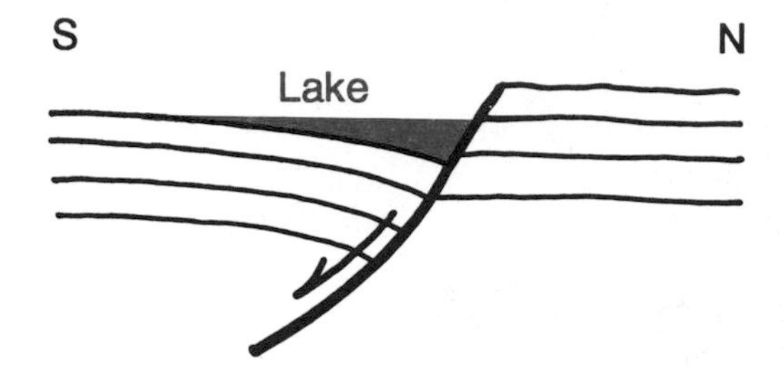

Lakes Pontchartrain and Maurepas may have formed in shallow depressions created by faults.

the city was built on the Pine Island trend of firm sand. The International Trade Mart and the Greater New Orleans River Bridge are on pilings driven into firm sand or stiff clay. These clays, which generally lie at depths between 60 and 100 feet, are considered bedrock.

The unconsolidated sediments beneath the city continue to subside because the delta is sinking under its own weight and because they are compacting, both natural processes. Under natural conditions, the river adds enough sediment to the top of the delta plain to compensate for its subsidence, but because the levees have channeled the sediment offshore for 300 years, large parts of the city and its environs are below sea level. How does the city stay dry? Pumps. The city pumps water up through the 17th Street canal to Lake Pontchartrain.

New Orleans stays dry during floods by diverting water from the river channel into floodways. The Bonnet Carre spillway, 30 miles upriver, was built in 1932 to divert floodwater to Lake Pontchartrain. The Morganza spillway was built above Baton Rouge in 1954 to divert floodwater into the Atchafalaya River basin, thus protecting both Baton Rouge and New Orleans.

Pontchartrain Causeway and Lake Pontchartrain

It is a long 24 miles across the Pontchartrain causeway. The water laps at the stanchions, seabirds fly past, and boats pass in the distance. The drive provides a good occasion for passengers to read this section.

Lakes Pontchartrain and Maurepas are north of New Orleans, sandwiched between the city and the low hills of the Prairie Complex that stretch along their north shores. Pass Manchac and North Pass connect them. The Amite, Tickfaw, Tchefuncte, Blind, and Tangipahoa Rivers supply fresh water from their drainages in the hills of the Florida Parishes.

Both lakes are large, but shallow: Lake Maurepas covers 93 square miles, and averages 7 feet deep. Lake Pontchartrain covers 635 square miles, and averages 12 feet deep. Bottom sediment is mostly silty clay, but patches

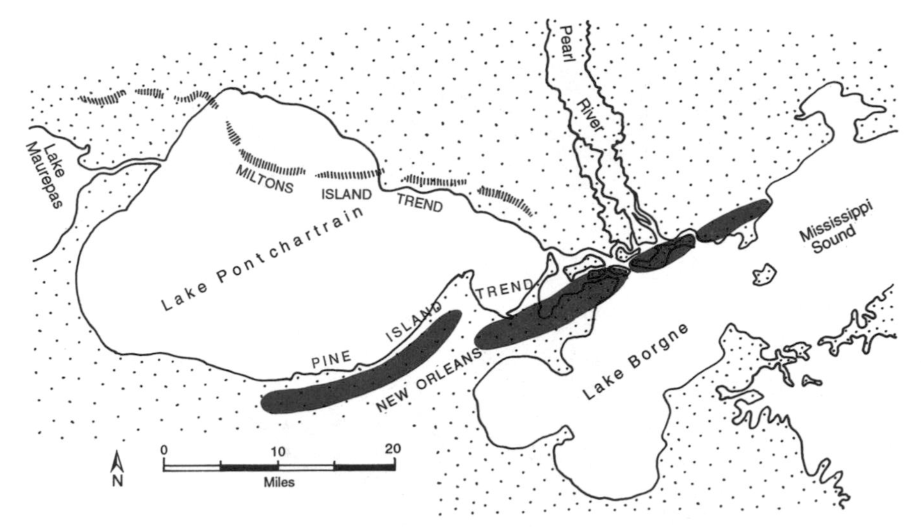

Map of sand shoals underpinning the area around New Orleans.
—Flowers and Ishphording, 1990; Otvos, 1978

of fine sand exist in the northern part of Lake Pontchartrain. It was probably reworked from old barrier islands.

The lakes rest on a complex of fairly recent lake, delta lobe, channel fill, and barrier island deposits, which add up to about 100 feet of sediment. These lie on Pleistocene deposits that correlate with the Prairie Complex sediments exposed a few miles north. The recent and Pleistocene sediments meet in a sharp line north of the lakes, along the Baton Rouge fault system, which offsets them about 50 feet. Some geologists suggest that the gentle slope into the fault of the dropped layers of Prairie Complex sediments may help explain the low area the lakes flood.

About 4,600 years ago, a train of barrier islands or shoals developed in shallow seawater southwest of a marshy mainland in what is now the New Orleans area. These sandy deposits were discovered in pits dug east of the city in the 1970s, when the Morrison Road ramp and Bullard Road were under construction. Geologists who studied the pits, as well as bore hole samples, recognized that the sands formed a line of barrier islands or shoals. They called them the Pine Island trend.

About the same time the sands of the Pine Island trend were deposited, the Mississippi River was building a delta eastward. It buried the Pine Island sands about 3,800 years ago, isolating a bay that would later become Lake Pontchartrain. A new line of barrier islands developed in the mouth of the open bay about 2,900 years ago, while another set of sandy shoals, the Milton Island trend, formed within it. About 300 years later, the St.

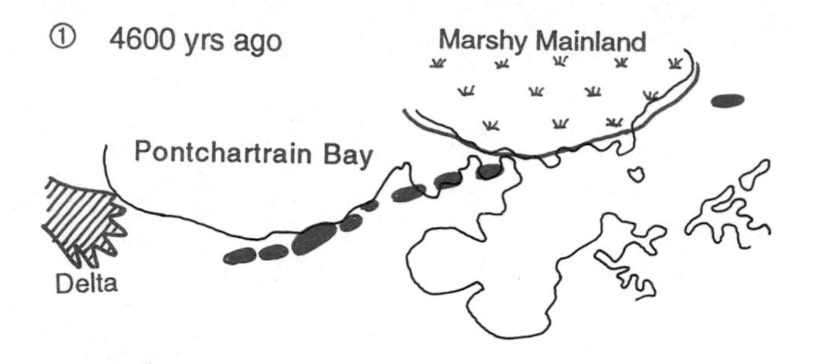

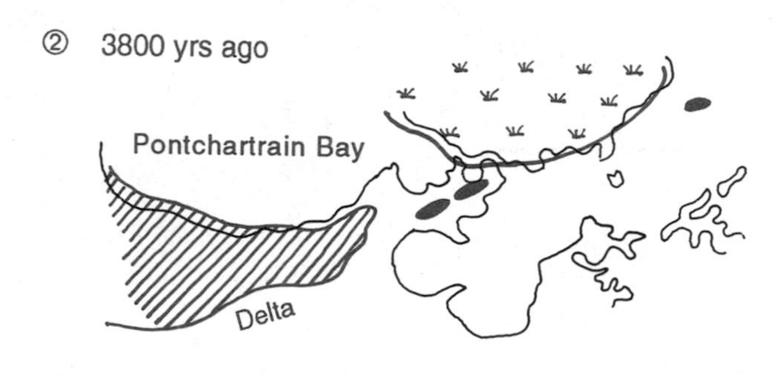

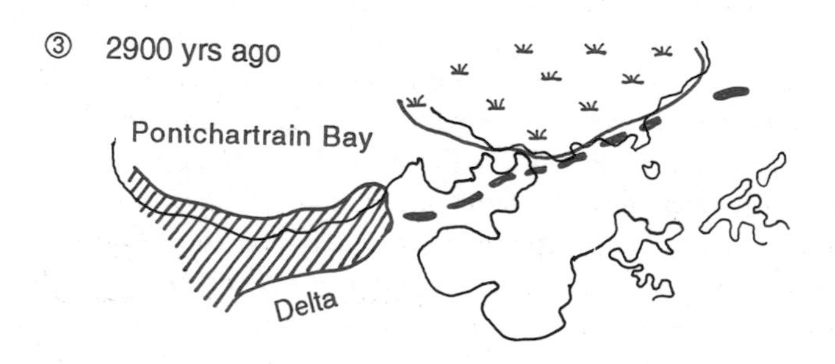

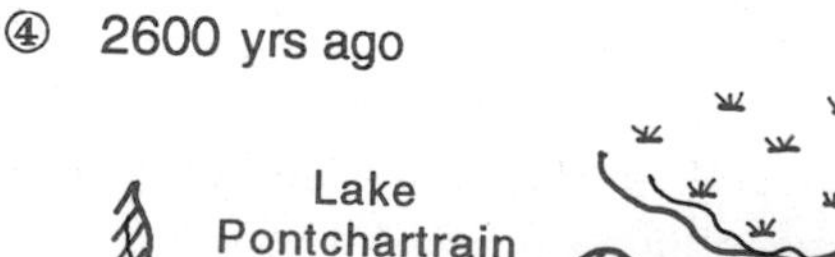

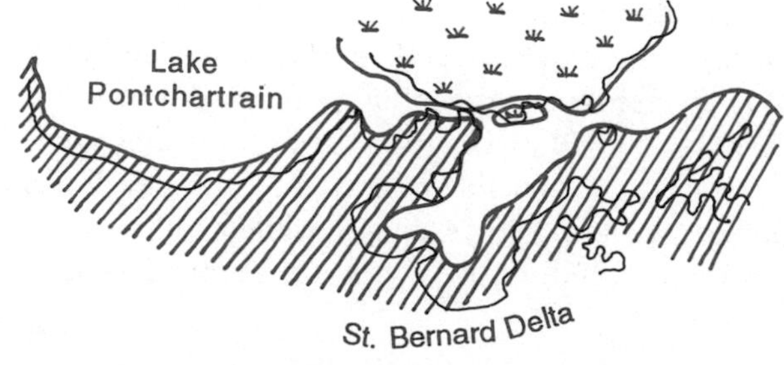

History of land formation around Lake Pontchartrain and New Orleans. —Otvos, 1978

Bernard delta of the Mississippi River had grown large enough to isolate the bay, converting it into Lake Pontchartrain.

Then Lake Pontchartrain expanded and Lake Maurepas appeared as the St. Bernard delta subsided. The Mississippi River switched to a new course, and abandoned the St. Bernard delta. Since then, the delta has been sinking, while waves and currents reshape its front into the arcuate string of the Chandeleur barrier islands, some 70 miles east of New Orleans.

Lake Pontchartrain has changed very little geologically in the past 2,200 years. But settlement of the lakeshores and the growth of New Orleans have brought great changes. Concerns for Lake Pontchartrain center on its proximity to the urbanized corridor of New Orleans. Pollutants are dumped into the lake from bridge, causeway, and highway drainage, storm water, spillage from boats, and general urban runoff. Upland runoff in streams from the Florida Parishes may also be a source of pollution. The Tangipahoa River supplies the lake with high fecal coliform concentrations from dairy farms, and sand and gravel mines produce excess sediment in surface water of the Amite River.

Heavy metals cause particular concern because they can contaminate fish and shellfish, poisoning the people who eat them. Sources of lead include exhaust emissions, solder, paint pigments, and pipe corrosion. Cadmium comes from pigments and electroplated metals; copper and zinc from pipe corrosion.

Some people have tried to label Lake Pontchartrain a dead lake. But it is far from dead. The clays in its bed help keep it alive by absorbing heavy metals and trapping some organic pollutants so they're unavailable to plants and animals. Shell dredging stirs up sediment, and could push heavy metals into the water, but they get tied up in the clay and soon settle to the bottom. Dredging has been curtailed, nevertheless, to help clear the lake.

Clay helps, but Lake Pontchartrain does have pollution problems. Bottom-dwelling organisms near outfall canals contain high concentrations of lead, silver, and cadmium, which probably came from repeated exposure to shock loads from urban runoff, not from bottom sediment. Clay also helps minimize environmental problems, as do monitoring and eliminating leaded auto fuels and paints, and disposal of metals.

Interstate 10
Baton Rouge—Lafayette
52 miles

This section of Interstate 10 crosses one of the great waterways and wilderness swamplands of North America. It crosses the Mississippi River and plunges deep into the adjoining Atchafalaya River basin, with its swamps full of trees shrouded in Spanish moss and its network of watery channels. The Atchafalaya River basin symbolizes the intricate balance of natural and artificial forces that shape so much of Louisiana's lowlands. This highway takes you past both the natural and engineering aspects of that balance, the factors that at once restrain and control this largely intact wilderness.

The high bridge over the Mississippi River provides magnificent views of the river, a mile wide and 160 feet deep. Ships anchor at Port Allen on the west bank to transfer their cargoes to barges, because the ships cannot pass under the bridge and because the water shallows upriver. Watch for the high levees that stretch along the riverbanks like gigantic green snakes. You can see that Baton Rouge is on the Prairie Complex east of the river. This is the first high ground you'll encounter while traveling up the Mississippi River, which explains why the city is here—230 river miles from the Gulf of Mexico. The lowlands west of the river support farms in the protection of the levees.

The Mississippi River levee and the East Atchafalaya protection levee shelter farmland between the Mississippi River and Ramah from floods.

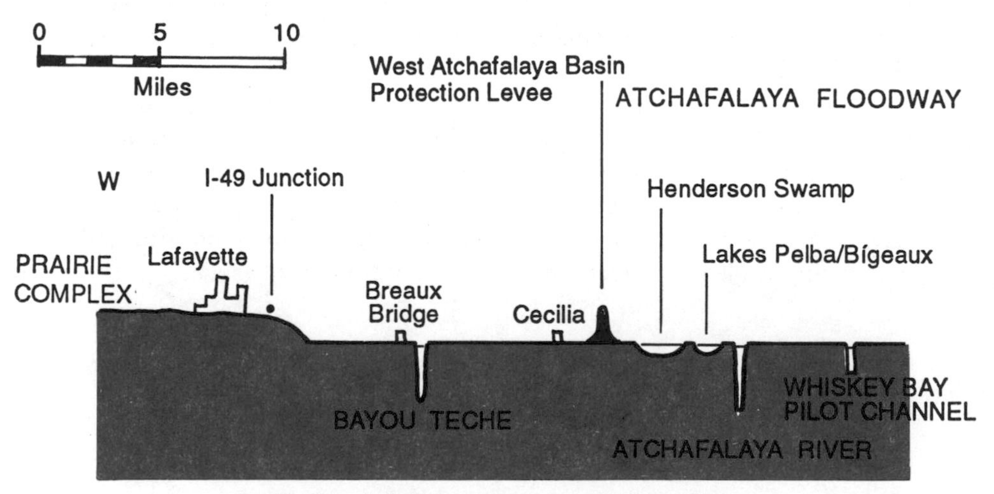

Profile along I-10 between Lafayette and Baton Rouge.

East Atchafalaya protection levee of Atchafalaya River basin. Along I-10, west of Ramah.

Look for the high East Atchafalaya protection levee running perpendicular to the highway on the west side of Bayou Maringouin, just west of Ramah. Bayou Maringouin is probably a crevasse distributary that overlies a former channel of the Mississippi River that flowed here between about 8,200 and 9,200 years ago.

In the nearly 20 miles between Ramah and Cecelia, the highway rides high over the Atchafalaya floodway on an elevated causeway. You see a magnificent view at treetop level of miles of tangled swamps punctuated here and there by open waterways.

The bridge over the Whiskey Bay pilot channel, eight miles west of Ramah, provides another high view. The channel is an artificially maintained waterway, the main navigation channel of the Atchafalaya River south to Six Mile Lake near Morgan City. The pilot channel now carries the

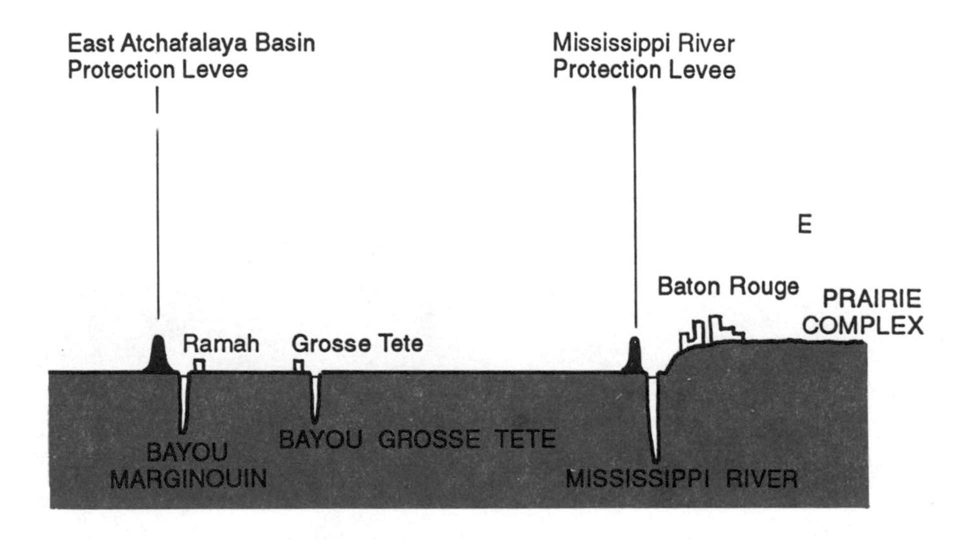

Atchafalaya River basin is crossed by elevated I-10.

main flow of the Atchafalaya River. Five miles west of the pilot channel, the road crosses the Atchafalaya River, originally the main natural channel, but now reduced to secondary flow status.

Lakes Pelba and Bigeaux are west of the Atchafalaya River, in the backswamp lowlands bordering the river. Watch for high-bridge views of Lake Pelba.

Innumerable stumps in Henderson Swamp tell of a magnificent forest of 600- to 700-year-old cypress trees that were cut for timber. Steamboats towed red cypress logs through Bayou Teche and the Atchafalaya River between the 1870s and the 1920s, the glory years of the cypress industry. Red cypress lumber is no longer available. The wooden platforms in the open water of Henderson Swamp are old oil and gas well locations.

The highway descends from causeway height to ground level where it crosses the West Atchafalaya protection levee, just west of Henderson

Henderson Swamp and elevated I-10. Wooden oil platforms and cypress stumps are remnants of former industries.

Swamp. It is securely above flood level, sandwiched between the natural levee and the even higher surface of the Prairie Complex. Watch a few miles east of Lafayette and Interstate 49 for the abrupt change in elevation as I-10 climbs out of the lowland of the Atchafalaya basin onto the Prairie surface on which Lafayette stands.

About halfway between the west levee and the crossing of Interstate 49, the highway crosses Bayou Teche, the legendary waterway of Cajun history and folklore. It was the main channel of the Mississippi River between about 3,800 and 5,500 years ago. Breaux Bridge, the crawfish capital, is on Bayou Teche just south of the freeway.

The undeveloped aspect of the Atchafalaya basin becomes clear through the scarcity of towns or side roads in the floodway land between the east and west protection levees. This contrasts sharply with the farms, roads, and buildings in the protected land outside the levees.

Interstate 10
New Orleans—Baton Rouge
67 miles

This route crosses modern swamplands around Lakes Pontchartrain and Maurepas, and higher ground of the Prairie Complex in the western half of the drive near Baton Rouge. The expressway stays a respectful five miles away from the Mississippi River, tracking its great curve as it bends from a north to south course at Baton Rouge to east to west at New Orleans.

Bonnet Carre spillway. Mississippi River is to the right of photo. Spillway releases floodwater to left into Lake Pontchartrain.

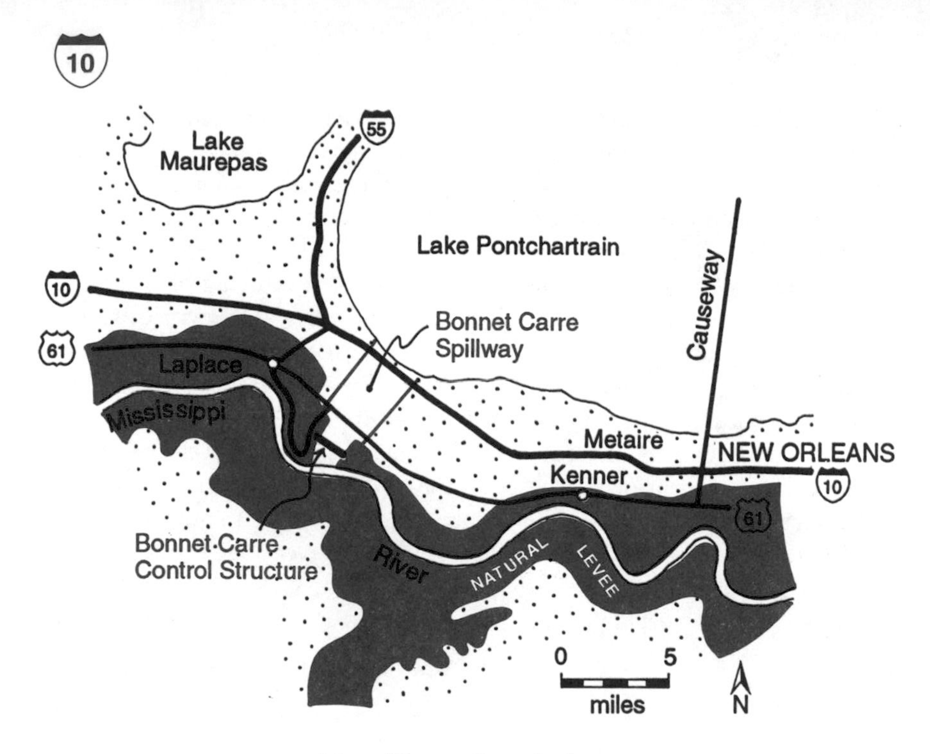

Map of Bonnet Carre floodway.

Sea level swampland bordering the south shore of Lake Pontchartrain has been largely "reclaimed," so the scenery west of the Pontchartrain causeway in Metairie and Kenner is mostly urban.

The highway crosses the Bonnet Carre floodway a few miles east of the I-55 junction. It is a broad area confined within levees that connects the Mississippi River and Lake Pontchartrain. The steel and concrete Bonnet Carre control structure at the Mississippi River end is opened in times of extreme flood to let water flow through the floodway and into Lake Pontchartrain. It then passes eastward through the Rigolets, a pass at the eastern end of Lake Pontchartrain, and into the Gulf of Mexico at Mississippi Sound.

The Bonnet Carre control structure is designed to divert water into the floodway before it tops the levees, thus minimizing property damage. It was built in 1936, and has been opened seven times since then, each time diverting floods from New Orleans. A short drive south on the River Road, Louisiana 628, takes you to the Bonnet Carre control structure, and to a view of the Mississippi River from the high control levees.

West of I-55, the highway crosses a bridge system through the middle of a wilderness of cypress and tupelo swamp. As you whiz along at highway speeds, try to imagine yourself paddling through the swamp in a wooden pirogue with an expert Cajun guide. The views of the unspoiled swamp from the elevated road are marvelous.

About where Louisiana 22 crosses the interstate highway, lowlands give way northward to higher ground, where pines and deciduous trees signal a change in the underlying rocks. Between this point and Baton Rouge, the highway is on the Prairie Complex. You can see occasional patches of tan silt, mud, or sand. The Prairie surface is about 50 feet above sea level and only slightly undulating. A few small streams empty into the Amite River, which flows to Lake Maurepas.

Interstate 10
New Orleans—Slidell—Mississippi
40 miles

Interstate 10 loops from the Lake Pontchartrain causeway eastward through New Orleans, around the south side of Lake Pontchartrain, to join Interstate 12 north of Slidell. In the city, the road is on silty and sandy alluvium of the Mississippi River. The alluvium is now mostly paved and covered with buildings for nearly 15 miles east of the Pontchartrain Causeway, about to the I-510 exit south to Chalmette.

In the more open country around the I-510 exit, the road crosses marshy land with cane and scrubby trees. East of Pte. aux Herbes, the road arches over six miles of open water, a bulbous bay on the east end of Lake Pontchartrain. The northern end of the bridge is on the firmer ground of the Prairie Complex, a few tens of feet above sea level. Interstate 10 runs north past Slidell for five miles, then makes a sharp turn east toward the Mississippi line.

Two miles east of the junction with I-12, and I-59, the highway drops off the surface of the Prairie Complex and enters the broad floodplain of the Pearl River, which is about four miles wide on the west side of the main channel. The swampy lowland, with waterways and cypress trees, contrasts strongly with the adjacent upland of the Prairie Complex.

Interstate 55
Hammond (I-12)—Interstate 10—Lake Maurepas
28 miles

In the five miles south of Hammond, I-55 crosses the Prairie Complex. A mile south of Ponchatoula, it abruptly descends as it crosses the Denham Springs–Scotlandville fault to modern alluvium at sea level. The fault extends in an east to west trend all the way to Baton Rouge. Its straight trace on the map suggests that it is nearly vertical. The south side dropped, helping create the lowlands that hold Lakes Maurepas and Pontchartrain, which are close to the fault, and on the side that dropped. The right-angle bend in the Tickfaw River, northwest of Lake Maurepas, is a surface expression of the Baton Rouge fault. It extends into Lake Pontchartrain where offsets are on the lake bottom.

The highway is elevated for miles where it follows the narrow ribbon of lowland cypress swamp that separates Lakes Maurepas and Pontchartrain. At North Pass and particularly from the high bridge over Pass Manchac, the two waterways connecting the two lakes, you can see the open water of Lake Maurepas on the horizon to the west. The elevated roadway provides excellent views of marsh and swampland. Watch for the cypress stumps on the south side of Pass Manchac. This narrow ribbon of land between the lakes is amazingly wild, with surprisingly little development, considering its proximity to New Orleans.

US 90
Lafayette—Morgan City
70 miles

Between Lafayette and Morgan City, Highway 90 makes a northwest-southeast arc, riding the west edge of the Atchafalaya basin along Bayou Teche, though for much of the distance staying high on the Prairie Complex. The Prairie Complex is between the Atchafalaya basin to the east and the fresh marshes that line the coastal bays to the west. Its surface is nearly flat to faintly undulating, and covered with loess. You can see occasional glimpses of its light tan silt or mud in road- or stream cuts.

Lafayette is on the Prairie Complex, above the flood level of Bayou Teche, but it is close enough to the Atchafalaya basin to take advantage of waterways for transport, and in the early days, to enjoy its rich resources of timber, fish, and crawfish. By the 1950s, Lafayette had grown into an important oil center. Its natural history museum and planetarium have displays on the geology and natural history of the area.

At the southern edge of Lafayette, the highway crosses the Vermilion River, which connects the Atchafalaya basin to Vermilion Bay to the southwest. It flows southwest; the other rivers in the Atchafalaya basin flow southeast. Near the airport, old ridges and swales that were part of ancestral Mississippi River meander belts make a gently rolling landscape on the Prairie surface.

Watch about five miles south of Lafayette for the exit to Louisiana 96, which goes east seven miles to St. Martinville. That is where the original

Acadians arrived from eastern Canada in 1765 to start their culture in Louisiana. The town was home to Emmaline Labiche, the heroine model for Henry Wadsworth Longfellow's epic poem, *Evangeline*.

From US 90, Louisiana 96 slices through the eroded and dissected edge of the Prairie Complex to the natural levee of Bayou Teche, on which St. Martinville stands. Bayou Teche is a former main course of the Atchafalaya River. It is within the Atchafalaya basin, but outside the main levee that defines the flowing Atchafalaya River system. St. Martinville is very historic and worth the short side trip.

Still on the Prairie terrace, the highway passes New Iberia. Established in 1779, it became a commercial center for rice and sugar cane production. Rich farmlands still surround the highway. West and southwest of New Iberia are the Five Islands: Jefferson, Avery, Weeks, Cote Blanche, and Belle Isle, strung in a line about five to ten miles from the highway. These circular mounds that tower 100 feet above the surrounding marshes are the tops of salt domes rising from the deep layer of Jurassic Louann salt.

The road follows the Prairie Complex past Jeanerette, on Bayou Teche north of US 90. It is the site of the Chitimacha Indian reservation, noted for its exquisite double-walled cane baskets, which are displayed at the tribal museum. The road crosses a last patch of Prairie Complex in the

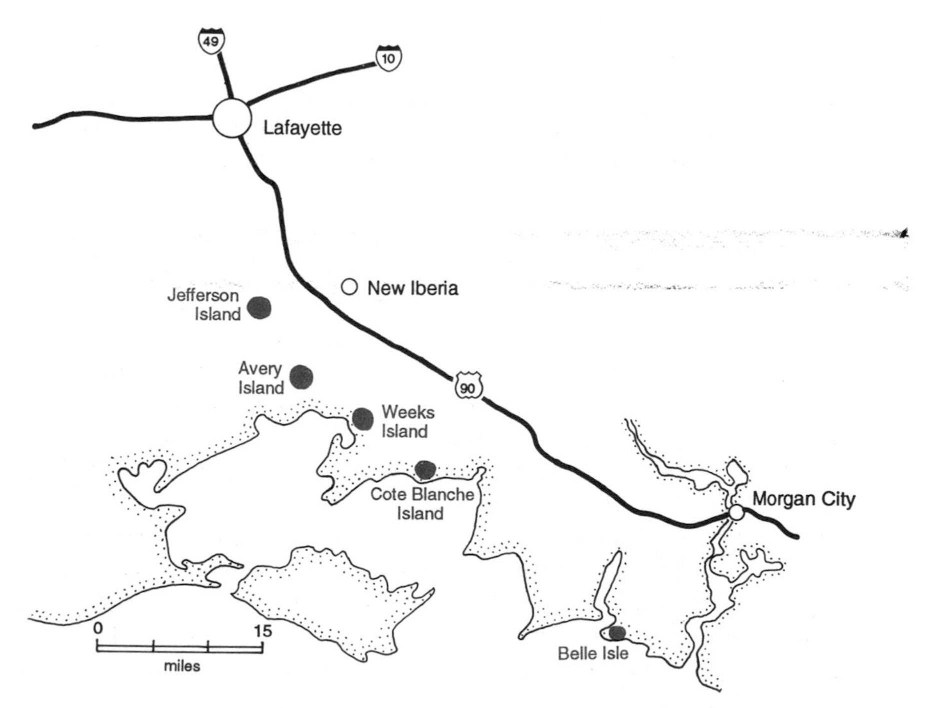

The Five Islands—surface mounds arched over salt domes.

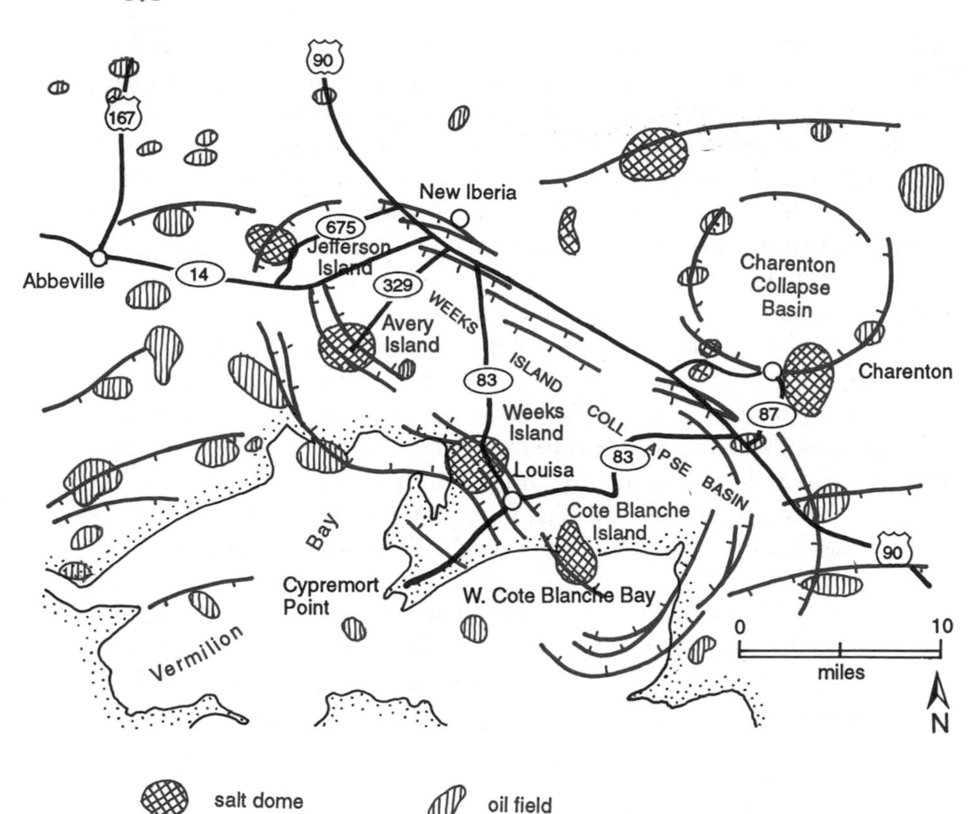

Collapse fault systems around the prominent salt "islands" near New Iberia. Note how salt domes are located around the edges of the collapse structures, but most oil fields are located near faults not associated with the collapse basins. —Seglund, 1974

area southeast of Baldwin, where the vegetation, landscape, and land use change dramatically. Between Baldwin and Morgan City, swamp and marsh barely above sea level fringe the highway. Delta plain and fresh marsh lie between US 90 and the coast, and the forested swamps of Bayou Teche and the Atchafalaya basin extend for many miles northeast of the road.

West of Calumet, about ten miles west of Morgan City, the highway crosses a large canal that connects Grand Lake in the Atchafalaya basin with Wax Lake, a drowned former river mouth that extends north from Atchafalaya Bay on the Gulf of Mexico.

A high bridge spans the main channel of the Lower Atchafalaya Waterway on the west side of Morgan City, providing a good view of the town, levees, and waterways. The port of Morgan City serves offshore oil production facilities and the shrimp industry. Levees entirely surround the city, making it a walled island in times of flood. Swamp Gardens features raised platforms over swamps and displays that tell how people have adapted to life in the Atchafalaya basin.

The Lower Atchafalaya River flows into Atchafalaya Bay 15 miles south of Morgan City, where the river is building a new delta. Belle Isle, the southernmost of the Five Islands, is on the marshy coast of Atchafalaya Bay, 16 miles southwest of Morgan City. No road goes there.

US 90
Morgan City—Houma
37 miles

The 15 miles of road between Morgan City and Gibson follow the natural levee of Bayou Black east of Morgan City, where beautiful cypress swamps border the highway. Between Amelia and Gibson, the highway departs from the levee to cut across the two limbs of a north loop of the bayou. A few miles west of Gibson, the road rejoins the levee. At Gibson, US 90 and Louisiana 20 split, and US 90 turns southeast to follow Bayou Black into Houma.

Cypress swamp along US 90 east of Morgan City. Duckweed, a floating plant, covers the water.

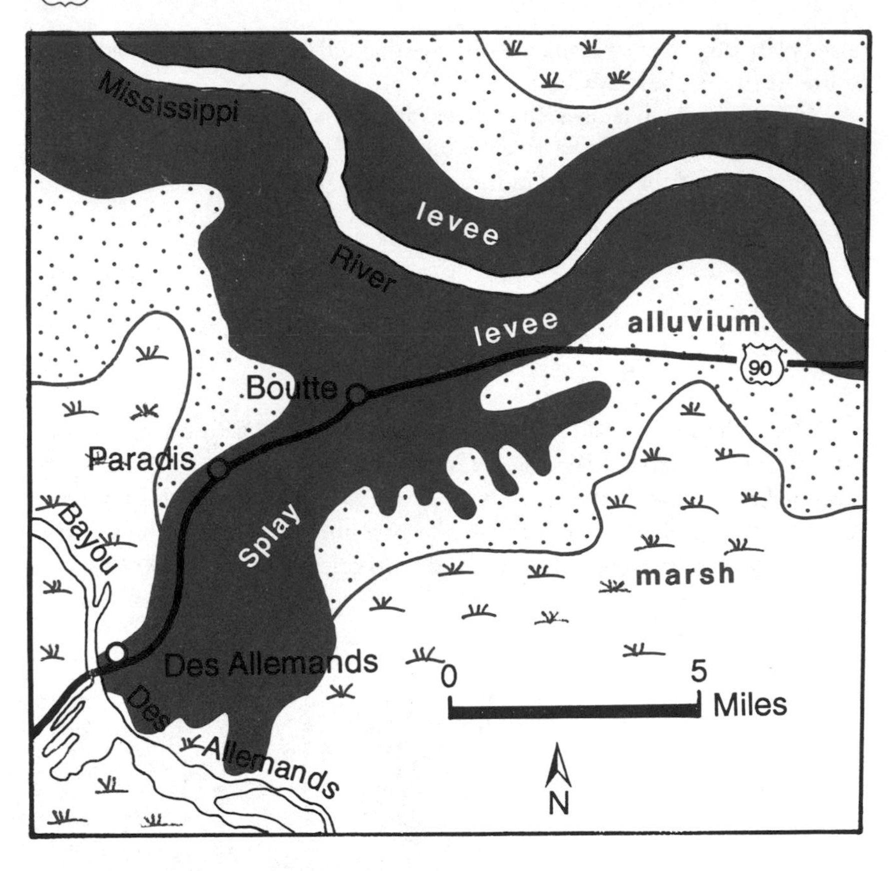

Crevasse splay from Mississippi River provides high ground for roads, towns, and agriculture in the Des Allemands area.

Bayou Black is a remnant distributary channel of the Lafourche delta, which the Mississippi River built between 2,500 and 800 years ago. A number of such channels around Houma flow generally southeast, toward Terrebonne and Timbalier Bays. But geologists think Bayou Black flowed northwest, from Houma toward Gibson, because its channels and levees are 1 mile wide and 14 feet high near Houma, and just a quarter mile wide and a foot high near Gibson. As the Lafourche delta built southeast to the Gulf of Mexico, the Bayou Black channel may have been a shortcut off the delta into the bay to the west.

Many bayous, the Houma navigation canal, and the Intracoastal Waterway all converge at Houma to make the town a bit of a Venice. Named after the Houmas Indians, Houma was founded in 1834 as the seat of Terrebonne Parish. For many years it has been a supply center for offshore oil and gas operations.

90

US 90

New Orleans—Des Allemands—Raceland—Houma

42 miles

You get a spectacular view of New Orleans where US 90 crosses the Mississippi River on the high Barataria Bridge. The highway passes through an urban landscape in the eastern ten miles of the route; the country farther west is more open and forested.

Boutte, Paradis, and Des Allemands stand on a large crevasse splay deposit of the Mississippi River. Des Allemands is ten miles from the river at the very end of the splay. The elevations of natural levees and crevasse splays, three to five feet above sea level, provide tolerably dry farmland.

West of Des Allemands, the road crosses Bayou des Allemands, which connects Lac des Allemands to the northwest and Lake Salvador to the southeast. Both are freshwater lakes in fresh marshes between the Bayou Lafourche and Mississippi River distributaries. The highway crosses about six miles of this fresh marsh southwest of Bayou des Allemands. The marsh vegetation is a tightly intergrown mat that floats on fresh water like a rubbery skin. Some geologists think floating marsh starts out growing on underlying clay, then tears loose in storms or hurricanes. Others believe floating marsh has always floated and invades new areas as water encroaches on it.

Three miles east of Raceland and the junction with Louisiana 1, the highway leaves the floating marsh to climb onto the higher natural levee of Bayou Lafourche. The highway bridges the main channel of Bayou Lafourche just east of Louisiana 1. It is hard to believe this small river was once the main channel of the Mississippi River, but it was just that from 2,500 to 800 years ago, while the river built its Lafourche delta.

Farther southwest, US 90 crosses about four miles of fresh marsh between river channels and levees on the Lafourche delta. The six miles of road northeast of Houma cut across Bayou Blue and Bayou Terrebonne, distributary channels of the old Lafourche delta. More than 800 years ago these small channels were huge rivers, like the channels on the Birdfoot delta of the Mississippi River south of New Orleans.

Louisiana 1
Port Allen—Donaldsonville—Thibodaux—Raceland
80 miles

This is a run along modern and ancient paths of the Mississippi River. Between Port Allen and Donaldsonville, Louisiana 1 crosses the wide natural levee of the modern Mississippi. The natural levee of Bayou Lafourche between Donaldsonville and Raceland is equally wide.

From 2,500 to 800 years ago, the main channel of the Mississippi River flowed down its old Bayou Lafourche channel. Then the river shifted its course east, leaving behind a natural levee five to seven miles wide and an abandoned channel now nearly filled with mud. All that remains is the trickle of Bayou Lafourche that you see next to the highway.

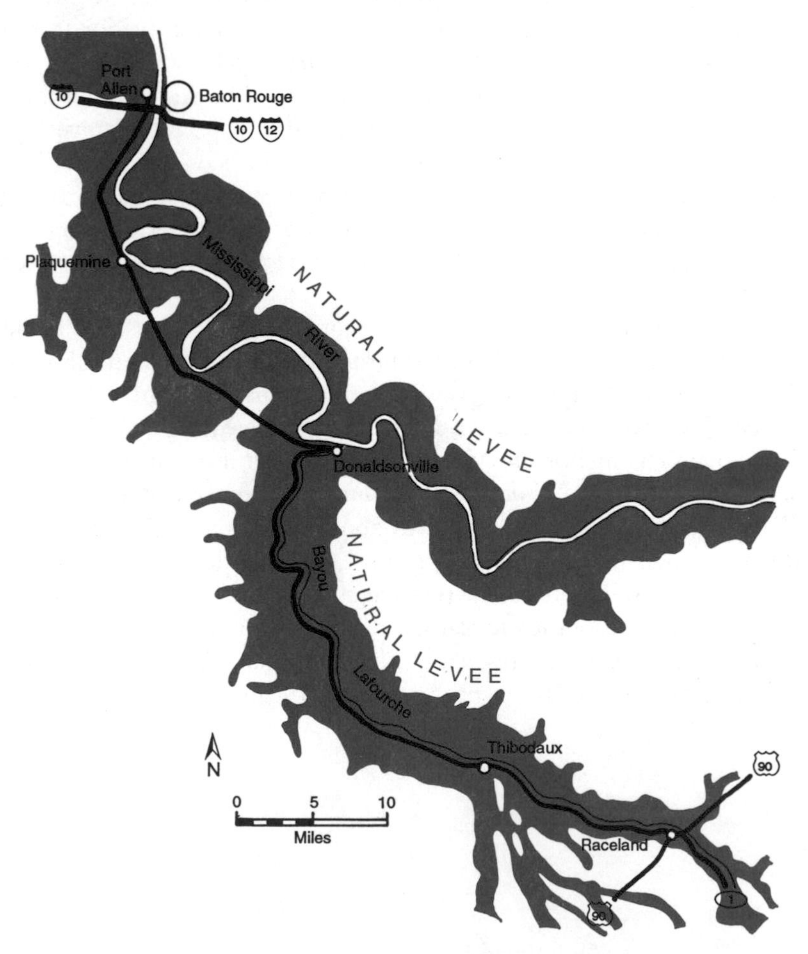

Natural levee deposits astride Bayou Lafourche are as wide as the levees bordering the Mississippi River, evidence that the Mississippi once ran down the Lafourche course.

Historical 1909 Plaquemine locks.

The Lafourche delta was also abandoned 800 years ago, and promptly sank beneath sea level because no sediment was deposited to offset the constant subsidence of the coast. Terrebonne Bay, Timbalier Bay, and Timbalier and East Timbalier Islands are all that remain of the old delta plain. As this area continues to sink, land disappears. Meanwhile, waves, hurricanes, and longshore currents erode the remnants of the Lafourche delta.

Thibodaux is at the apex of the old Lafourche delta. The map shows how the levee and channel system divides into distributary channels that fan southward. North of Thibodaux, the road follows the single Lafourche channel, formerly the Mississippi River. The broad farms and old plantation houses that line the way are on the old natural levee. Some of Louisiana's finest old plantation houses are near Thibodaux. Laurel Valley plantation on Louisiana 308 south of the city is touted as the largest turn-of-the-century sugar plantation in the south. Acadia plantation, on Louisiana 1 near town, was built by Jim Bowie in 1828.

Donaldsonville is at the junction of the Mississippi River and Bayou Lafourche, an important spot in 1750 when the town began as a trading post. The city is the third oldest settlement in Louisiana, after Natchitoches (pronounced *NACK-o-tish)* (1714) and New Orleans (1718).

The State Commemorative Area at Plaquemine preserves the 1909 locks that connected Bayou Plaquemine and the Mississippi River. Displays in the lockhouse explain river history, boats, and trade.

Alternate Louisiana 1 follows a beeline north along the Intracoastal Waterway from Morgan City to Port Allen. The Port Allen locks south of town transfer ships between the waterway and the Mississippi River. Dry-product wharves line the Port Allen shore, smaller docks and a large tanker terminal are on the Baton Rouge side, and several midstream dock-

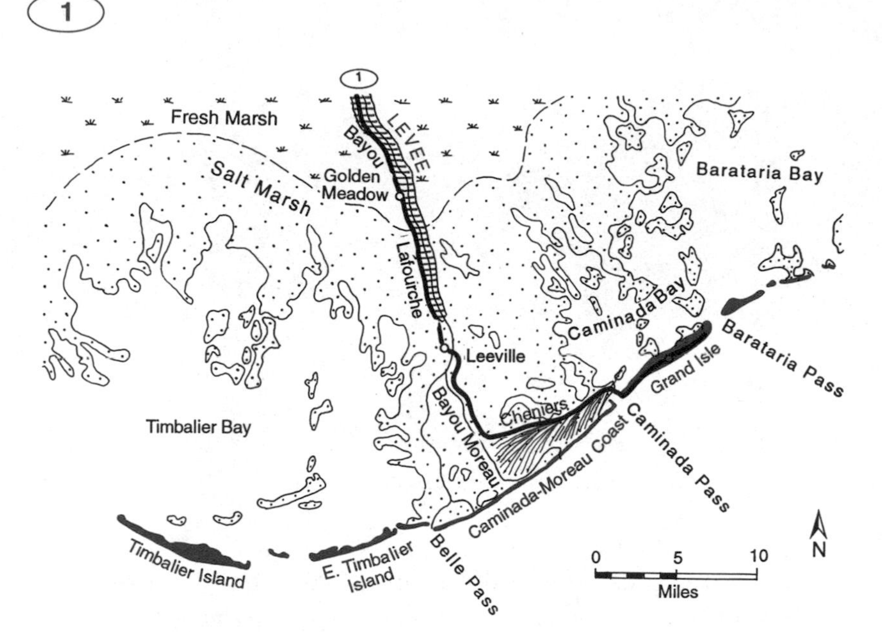

Geologic map of the Caminada-Moreau coast and Grand Isle area, showing the constructive and destructive environments of the Lafourche delta. —Penland et al., 1986

ing buoys transfer goods between ships and barges. In tonnage, the port of Baton Rouge is Louisiana's second largest, the fifth largest seaport in the United States, and the farthest inland deepwater port. Ships can navigate the Mississippi River to Port Allen, where they transfer their cargoes to river barges.

Louisiana 1
Raceland—Golden Meadow—Grand Isle
63 miles

Along most of this drive, the road stays close to Bayou Lafourche, a small stream that makes its way southeastward across a vast marshy plain to the Gulf of Mexico. Bayou Lafourche is the abandoned channel of a former course of the Mississippi River, which laid down the impressive network of levees, alluvial deposits, and extensive muddy marshes that surround Louisiana 1 near Raceland, Lockport, and LaRose.

From 2,500 to 800 years ago, the mighty Mississippi rolled where Bayou Lafourche now dawdles. The marshy delta plain east of Timbalier and Terrebonne Bays is a remnant of the delta it then built. After the Missis-

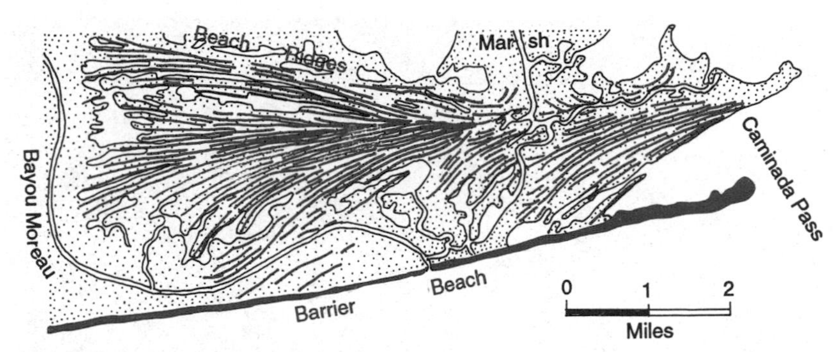

Close-up diagram of the fan of beach ridges behind the modern barrier beach on the Caminada-Moreau coast. —Penland et al., 1986

sippi River abandoned the Lafourche course about 800 years ago, the delta, deprived of its supply of sediment, compacted and sank—Timbalier and Terrebonne Bays are sunken portions of this old delta. Timbalier and East Timbalier Islands and Grand Isle are sandy barrier islands that the waves built as they reworked the front of the abandoned delta.

Sediment has nearly filled the old channel, leaving Bayou Lafourche to drain local water to the Gulf of Mexico. The two old levees are a mile wide at Raceland, and narrow gulfward. They are less than a half mile wide near Golden Meadow, and finally come to their seaward tips two miles north of Leeville. River alluvium borders the levees as far as Golden Meadow, then marsh meets them at Leeville. It is easy to spot the levees along Louisiana 1, because houses, buildings, and roads were built on them to take advantage of their elevation. You can see the decrease in the number of side streets that parallel the highway as you travel south, down the narrowing levees.

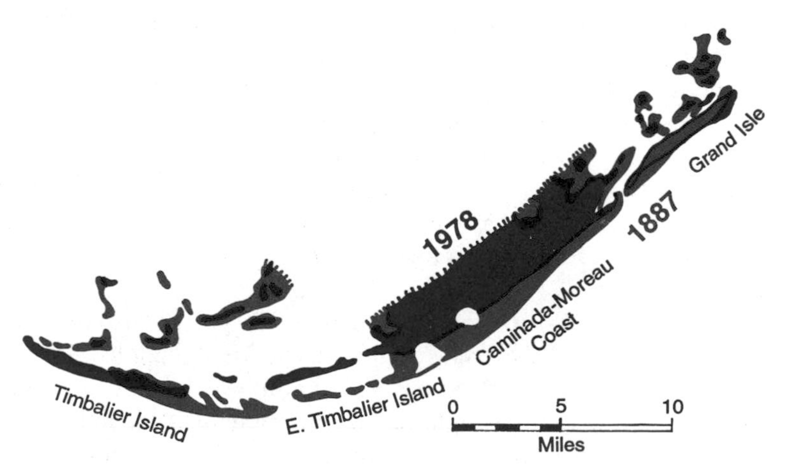

Map showing the dynamic change in the coastline southwest of Grand Isle during the past 100 years. —Penland et al., 1986

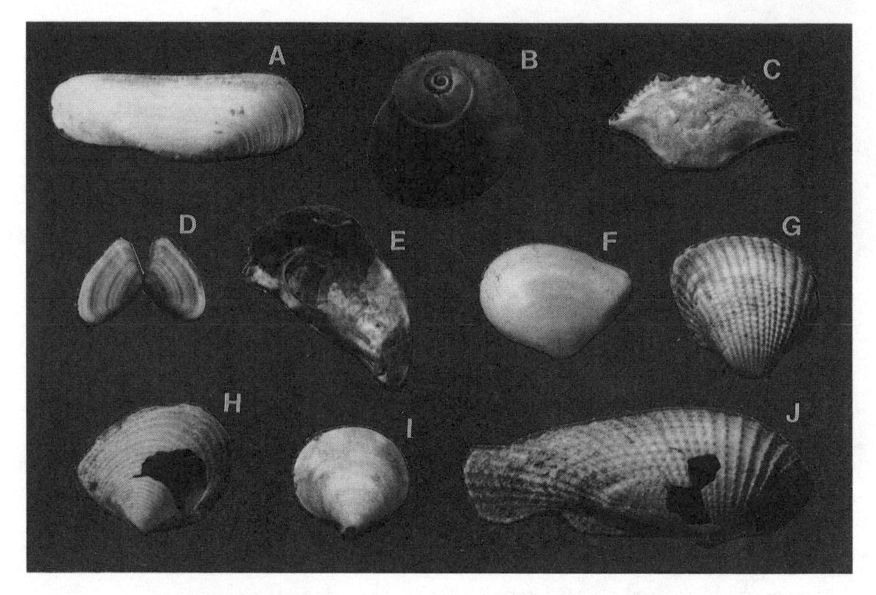

A	Tagelus clam	F	Venus clam
B	Moon shell snail	G	Cockle clam
C	Crab carapace	H	Channeled duck clam
D	Coquina clam (colorful bands)	I	Mussel shell
E	Oyster shell	J	Angel wing

Common shells found on Grand Isle beaches. Many are worn and broken by wave action in the surf zone just off the beach.

The fresh marsh that borders Bayou Lafourche yields to an immense salt marsh at Golden Meadow, which continues to the coast. Marsh grass and patches of open waterway stretch to the horizon from the high bridge at Leeville. The road continues across marshland in the seven miles south of Leeville, then turns east toward Grand Isle.

The ten miles of highway to the bridge at Caminada Pass travel between marsh to the north and an expanse of forested sand ridges known as Cheniere Caminada to the south. This, the largest beach ridge plain in the state, stretches south three miles to the Caminada-Moreau coast. It contains as many as 70 nearly parallel sandy beach ridges. Live oak trees grow on the ridges; the marsh grass, *Spartina*, grows in the watery swales between them.

These beach ridges began to grow about 700 years ago as distributary streams from the old Lafourche channel of the Mississippi River marched seaward over the older Terrebonne delta shoreline, which by this time was retreating. The streams brought in a large supply of sand, and so did the waves sweeping it west from the beaches on the old Terrebonne shoreline.

Observation tower at Grand Isle State Park. Vegetated dunes in foreground.

All that sand made the Cheniere Caminada beach ridge plain grow quickly and dramatically. About 300 years ago, the Mississippi River abandoned its Lafourche channel, greatly diminishing the supply of sand. For the past few hundred years, the Caminada-Moreau coast has retreated under wave attack. Waves focus on that stretch of coast because it is a headland that sticks out into the Gulf of Mexico.

Historical records and maps from the 1800s show how much of the Caminada-Moreau coast has been eroded. Meanwhile, waves and currents have moved the eroded sand to East Timbalier Island and Grand Isle, both of which have grown in the recent past.

At the east end of Cheniere Caminada, Louisiana 1 hooks south to cross the bridge between the Caminada-Moreau coast and Grand Isle. The water gap is Caminada Pass, a tidal inlet that connects Caminada Bay to the Gulf of Mexico. Tidal inlets are openings through coastal barrier beaches or a string of sandy barriers. Their main function is to let water drain from the land to the sea. They also let water flow into or out of the bay as the tides rise and fall.

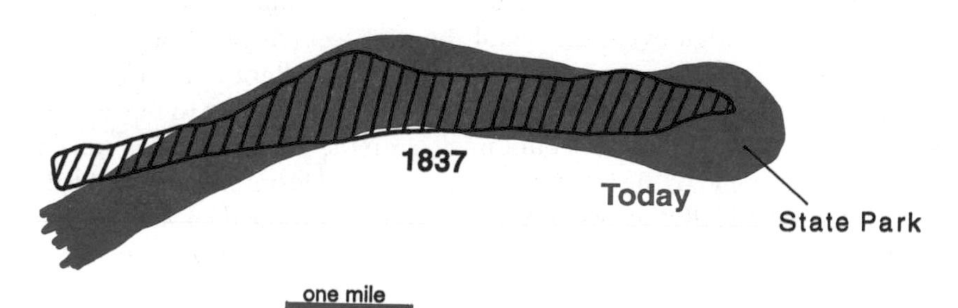

Map comparing Grand Isle in 1837 with its form today.

Plants on beach act as baffles to trap the windblown sand.

Grand Isle attracts thousands of people who come to enjoy its beaches, water, and dunes. It is a classic barrier island that separates the Gulf of Mexico from Caminada Bay and the larger Barataria Bay. And it is the easternmost bastion of the Bayou Lafourche shoreline.

Drive the six miles of the island to Grand Isle State Park, on the shore of Barataria Pass. From the observation tower, you can look across the Pass to Grand Terre, where the privateer Jean Lafitte had a base, and where the ruins of Fort Livingston guard the entry to Barataria Bay. It is Louisiana's one coastal fort directly on the Gulf of Mexico, built in the 1800s to control access to Barataria Bay. The walls are made of cemented clamshells, faced with brick and trimmed with granite. It was incomplete when the Civil War broke out, and never saw military action. A hurricane destroyed the south wall in 1915.

Grand Isle is a dynamic boundary between land and sea. Barrier islands protect the marshes from the waves, and host countless shore- and water birds. A survey map drawn in 1837 shows no land where Grand Isle State Park now is. That part of the island has grown since then as the waves import sand from the Caminada-Moreau coast.

Waves have naturally eroded the west end of Grand Isle; they deposit sand on its east end, moving the island about 16 feet per year. People have influenced this natural process. A 1958 jetty built at Barataria Pass at the east end of the island in 1958 accelerated beach growth there to more than 30 feet per year. A second jetty, built in 1972 on the west side of Caminada Pass at the other end of the island halted erosion and fostered a little growth on west Grand Isle. But growth on east Grand Isle dropped back to 16 feet per year.

Look for wind ripples on dune edges and on open beach flats. Watch sand grains bounce along before the wind. Find the insect and crab tracks marching boldly across the dunes. See how vegetation traps sand and stabilizes dunes, and how breaking waves churn up bottom sand, moving it shoreward. Watch how the swash and backswash of waves approaching at an angle wash sand diagonally up the beach, then carry it straight down the beach. This is what drives the longshore drift of sand along the beach. See if you can spot old beach ridges behind the dunes in the interior of the island. Compare the features of the side facing the gulf with those on the bay side.

Louisiana 14, Louisiana 329

Avery Island

6 miles

Avery Island.

Avery Island is the most famous of the Five Island salt domes, but perhaps more culturally than geologically. Millions of people around the world have read of Avery Island, Louisiana, on the label of a bottle of Tabasco sauce. This red pepper sauce has been made on Avery Island since 1859. The high ground, chert gravel for making stone tools, and salt springs drew Indians to Avery Island; white settlers followed.

Brine springs were discovered on the island as early as 1791. John Marsh bought the north half of the island in 1818 and produced salt by evaporating brine. Nineteen years later, Judge Daniel Avery married Marsh's daughter and eventually acquired the entire island. Their son, John, started brine evaporation during the Civil War. Pure rock salt was discovered 16 feet below the surface; in 1867, a salt mine went 53 feet into rock salt, but closed 24 years later because of water seepage and sinkholes, such as Blue Pond. The Avery family later dug a deeper mine, which still produces from a 500-foot shaft.

Just before the Civil War, Edmund McIlhenny married Judge Avery's daughter. Starting with pepper seed he got from Mexico, McIlhenny grew peppers and concocted a sauce from his peppers, Avery Island salt, and vinegar, which he aged in wooden casks. In 1868 his Tabasco sauce was trademarked, and a Louisiana culinary legend began.

Hilltop of Avery Island on skyline behind Tabasco factory.

Like a ghost ship floating a murky sea, Avery Island rises from the surrounding marsh, growing on the horizon as you approach from the northeast along Louisiana 329. It rises 150 feet above the wetland plain. It is easy to understand when it is wreathed in mist why early settlers called it an island.

Today we know that Avery Island stands above a salt dome, which rose thousands of feet through layers of lighter sedimentary rocks as though it were a hot air balloon rising through layers of air. The heavy load of sediments deposited above the Louann salt during Tertiary time squeezed it up in tall pillars, like long columns of oil rising through water. The Weeks Island collapse fault system promoted salt movement in this and three of the four other salt islands in this area. Most of the Avery Island salt dome rose between Eocene and Miocene time, but some movement probably continues. And groundwater is still dissolving salt at depth.

Avery Island is nearly circular, and covers about 2,500 acres. Blocky windblown dust, loess, lying on sediments of the Prairie Complex mantle its surface. Older geologic maps show these as Intermediate Terrace sediments, but recent research identifies the sands with the Prairie Complex. Streams have gullied the mound, and several ponds lie in surface depressions. The depth to salt ranges from 16 to 300 feet. Thousands of herons and egrets still roost in a swamp amid oil tanks, pepper fields, and salt mine workings.

Rapid growth of the lower Mississippi delta. —Scruton, 1960

Louisiana 23
New Orleans—Venice—Tidewater
77 miles

Going down the delta—the idea has an allure, a magic about it. To drive Louisiana 23 along the Mississippi River as it makes its final dash to the sea is a kind of roadside culmination—a conclusive wrap-up of all that is Louisiana geology.

The amazing thing about the modern Mississippi delta is how young and dynamic it is. Geologists emphasize the great ages of earth events, and the slowness of natural processes. Here is a major geologic feature that started to form only a few hundred years ago, and is changing from one day to the next.

Swamp on side road to Lake Hermitage.

The Mississippi River began to build its modern delta, which geologists call the Balize or Birdfoot delta, about 600 years ago. Early settlers built the first artificial levees along it in the late 1600s, and people have been interacting with it for half its life.

From US 90 in Gretna, Louisiana 23 heads southeast toward Belle Chasse through urban sprawl on the high ground of the natural levee, which is about five miles wide. At Belle Chasse, the road turns sharply southwest to parallel the Mississippi River all the way to Venice.

The immense artificial levee on the west bank of the river dominates the horizon east of the highway; farmlands and woodlands stretch away to the west, on the natural levee.

About five miles south of Belle Chasse, the natural levees narrow to strips about a mile wide on either side of the river. The river has not flowed through this channel long enough to build wider levees. And artificial levees have prevented natural flooding, restricting the amount of mud the river can spread across its natural levees.

If you look east between 10 and 15 miles south of Belle Chasse, you can see the low slope of the natural levee in open fields. To the west, the natural levee slopes gently downward from the highway. It is subtle, but real. Watch, as you travel south, how the marsh edges ever closer to the road, which reflects the progressive seaward narrowing of the natural levee on which the road is built. The scene hardly changes in 30 miles: artificial levee to the east, narrow natural levee to the west; marsh and swamp vegetation interweaving with slight variations in elevation, water depth, and water saltiness. If you drive up on the levee at Port Sulphur, you can look out over the river at the large ships. This is also an opportunity to see the river running within levees at a level above that of the road.

At milepost 46, watch for the side road to Lake Hermitage. This is a beautiful drive through five miles of untouched swamp and salt marsh. Garlands of Spanish moss festoon live oaks and cypress trees, and palmettos grow beneath them. Beyond the swamp, the waving grasses in the salt marsh reach to the horizon. Ghostly tree skeletons on the swamp fringe tell of sinking land and saltwater incursion.

The high bridge south of Empire affords an aerial view of lakes and marshes typical of the filled bays near the end of the delta.

Fort Jackson, between Triumph and Boothville, was built between 1822 and 1832 to guard the entrance to the Mississippi River. It is on high ground at the end of the natural levee. The fort was surrendered after the Union Navy shelled it for ten days after New Orleans fell in 1862. Fort Jackson was used as a training base in 1898, 1917, and 1918; it became a national monument in 1961.

Fort Jackson stands at the head of the fresh marsh that covers the entire end of the Birdfoot delta. The strong flow from the river spreads across

Old Fort Jackson.

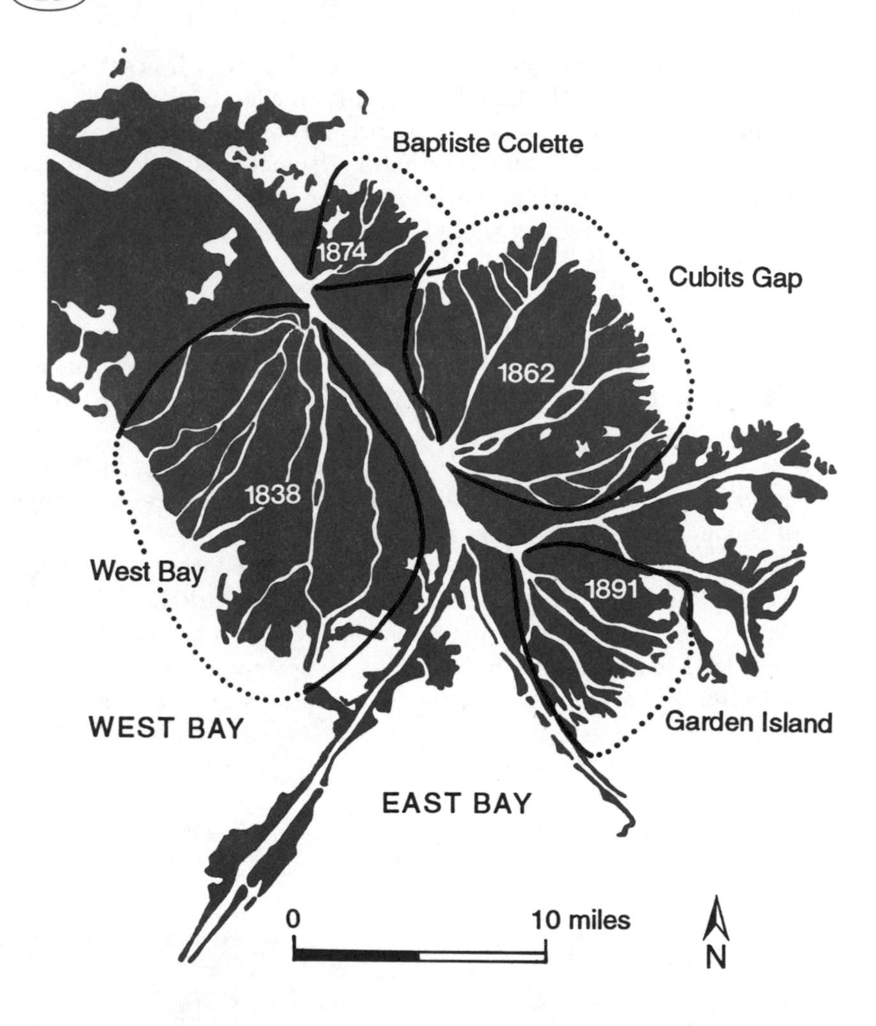

Pattern of bay filling by crevasse splays on the lower Mississippi delta. —Coleman, 1988

the marsh, pushing the seawater from this low land. River water is less dense than seawater, so it floats over the denser seawater to maintain the fresh marsh at the end of the delta.

The birdfoot pattern of the delta is quite noticeable seaward of Fort Jackson, where the river splays into a complex system of distributaries. Old maps record when various lobes of the birdfoot developed. The delta was skinny and clawlike in the mid 1800s, but bays between the toes of the birdfoot have rapidly filled with sediment to give it a fatter appearance.

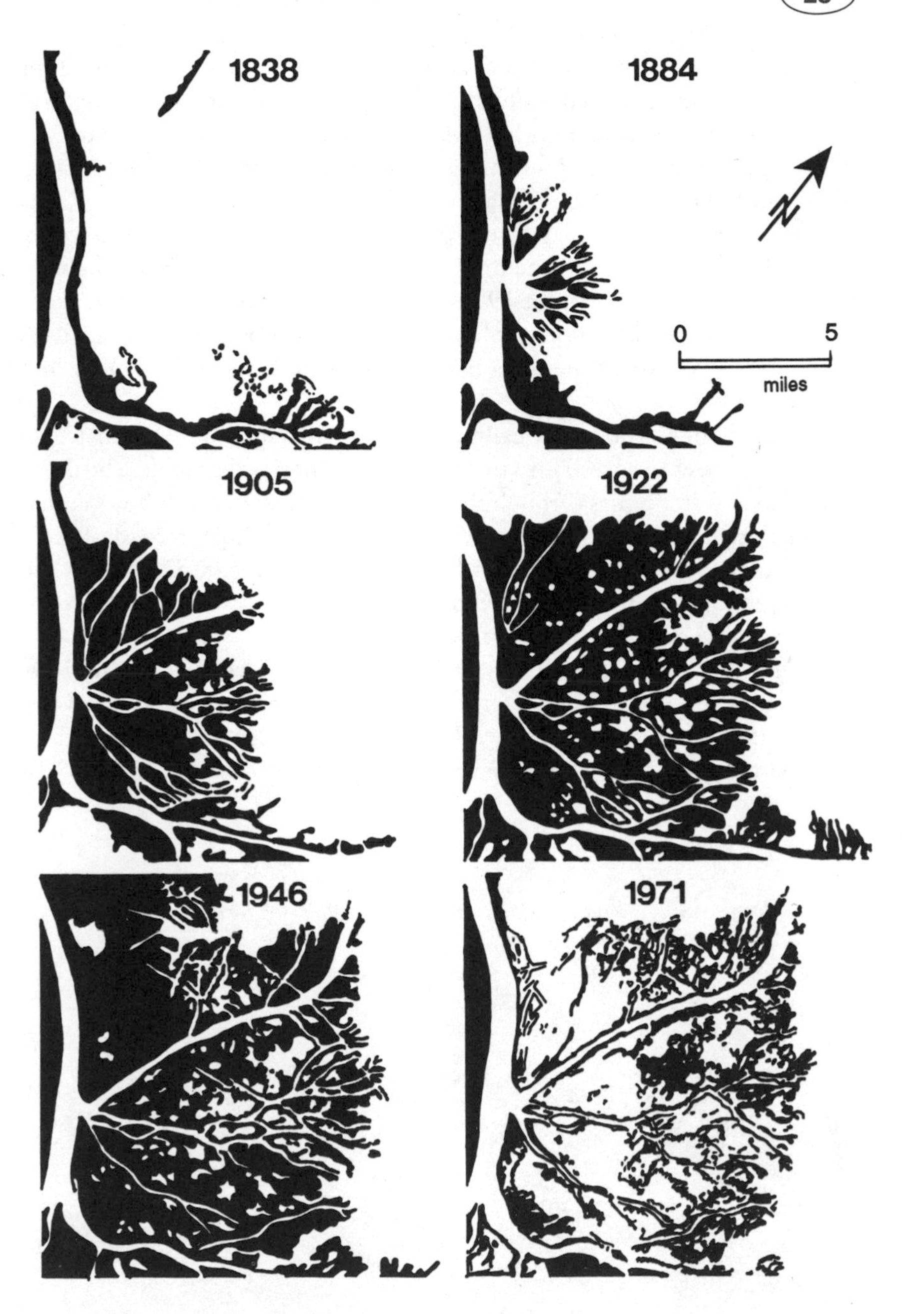

Birth, growth, and destruction of Cubits Gap bay fill. —Coleman, 1988

The history of one bay, Cubits Gap, is especially well documented. In 1862, a fisherman named Cubits dug a narrow ditch across the natural levee to make a shortcut to his oyster grounds. A flood later that year widened the ditch, which grew into a broad channel by 1868. Muddy river water poured through Cubits Gap, and filled the bay with sediment. Over the next 100 years, the bay filled, then subsided as the sediments compacted; by 1971, almost no channels were building in Cubits Bay. The map shows a lacy appearance in the 1980s because 75 percent of the constructed land in the bay had been lost to compaction and subsidence. Other bays on the birdfoot part of the delta have gone through cycles similar to that at Cubits Gap.

High floods erode cuts called crevasses through the natural levee, helping the river to spread into the many distributaries that build a birdfoot delta. River water pours through the gap, then loses speed as it flows across the shallow floor of the bay. As the water slows, it dumps its load of sediment, the heavier sand grains settling first, then the silt, and finally the mud. Geologists call the package of sediments a crevasse splay. Many of them coalesce and add to natural levees.

The river drops its sediment load at its mouth, where its flow slows and spreads because it is no longer confined within channel walls and levees. The larger particles settle first, while the finer particles spread seaward, so it is common to find a sandbar at the mouth of a river. As the river mouth

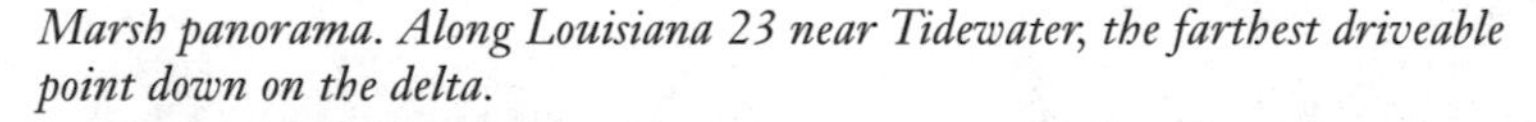

Marsh panorama. Along Louisiana 23 near Tidewater, the farthest driveable point down on the delta.

bar grows, it forms a bulbous barrier that eventually splits the river into two streams. This process, repeated again and again, creates a spreading delta made of bifurcating channels at the river's mouth.

The road continues to Venice and Tidewater, which are logistics and supply points for offshore oil and gas drilling and production rigs. They are also jumping-off points for boating and recreation on the delta. Between Venice and Tidewater, at the extreme end of the road, you can see excellent freshwater marshes, many water birds, waterways choked with water hyacinths, and trees on feeble levees. You feel immersed in the ephemeral environment of the delta.

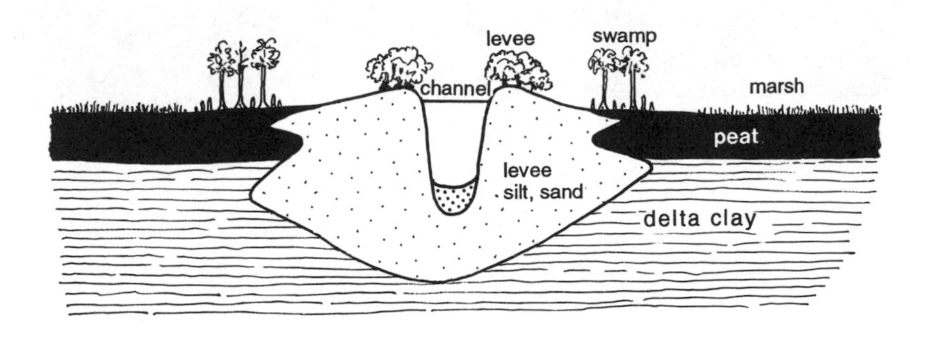

Profile across a distributary river channel on a delta.

Salt mine facilities of Morton International Salt, Weeks Island.

Louisiana 83, Louisiana 319
Weeks Island—Cypremort State Park—Cypremort Point
21 miles

This is a drive down the axis of a former main channel and delta of the Mississippi River, with a spectacular salt mound thrown in for good measure. You get great views of both freshwater and saltwater marshes bordering Cypremort Point, which separates Vermilion and West Cote Blanche Bays. Turn off US 90 onto Louisiana 14 toward New Iberia, drive about two miles, and take Louisiana 83 south to Cypremort Point.

In the northernmost five miles, the road stays on the solid ground of the Prairie Complex. About one mile south of the junction with Louisiana 85, it drops to freshwater marshland, and crosses it in the next five miles. Waving grasses and low shrubs dominate the scene until the sharply delineated slopes of Weeks Island rise abruptly, the top standing 170 feet above sea level.

Weeks Island is another of the salt dome beads on the Five Island chain, and is part of the Weeks Island collapse structure. As in the other islands, salt pushes up beneath Weeks Island from deep beneath the surface, from the Louann salt. It is so close to the surface that it was mined during the nineteenth century. Weeks Island still hosts a salt mine and chemical plant, as well as natural gas field and processing plant, and it contains a Department of Energy underground storage site for strategic petroleum reserves.

Erosion has gullied and dissected the Prairie Complex sediments and overlying loess on the island. Many of the gullies are remarkably straight, which suggests that they probably follow fractures and faults. These may reflect a broken salt surface below. The raised terrace sediments and the apron of sediment around the edge of the island suggest that the salt dome has been rising during the past few thousand years. It is probably still rising.

River sediments and freshwater marsh surround the road between Weeks Island and Louisa, where you may see Cote Blanche Island to the southeast.

Follow Louisiana 319 from Louisa southwest to Cypremort Point. *Cypremort* means dead cypress in Cajun French. The road is on natural levee deposits of Bayou Cypremort, five to ten feet above marsh level. Fresh marsh, seen through the trees, spreads out on either side of the road along the first half of the trip, then gives way to salt marsh in the last several miles. Cypremort Point State Park has a nice beach on the Vermilion Bay side of the point. This is the only beach you can drive to between the barrier island beach at Grand Isle and the chenier beaches near Cameron. Salt marsh stretches east from the marina toward West Cote Blanche Bay.

Bayou Cypremort is a small remnant channel of a former delta of the Mississippi River. From 7,500 to 5,000 years ago, the Mississippi River was far west of its present course, running down Bayou Maringouin and Bayou Cypremort to build a delta at Vermilion Bay and West Cote Blanche Bay. About 5,000 years ago, the river shifted east, and the Maringouin-Cypremort delta, starved of its supply of mud, began to sink. The marshes

Cypremort Point.

around Vermilion and West Cote Blanche Bays and Marsh Island on the gulf side of the two bays are all that remain of the old Maringouin-Cypremort delta.

Louisiana 675
Jefferson Island
5 miles

Louisiana 675 is a paved lane that turns off Highway 90 two miles north of Louisiana 14. The road wends its way across relatively flat Prairie Complex deposits, a clay plain with a cover of nearly three feet of wind-blown silt, called loess, which once supported rich cotton and sugar cane plantations.

Jefferson Island rises some 50 feet above the surrounding plain. It was known as Orange Island until Joseph Jefferson (1829–1905), a famous actor and artist, bought it after the Civil War. He built Jefferson Manor in about 1870. You can tour the greenhouses and gardens, and perhaps see an oil pump producing from sandstone tipped up against the flank of the Jefferson salt dome.

Jefferson Island is the northernmost of the Five Islands chain of salt domes. The salt pushed up through thousands of feet of sediments from its source in the deeply buried Louann salt, which was deposited during Jurassic time. A spine of salt under the island comes to within 100 feet of the surface. A larger salt surface lies about 869 feet beneath Lake Peigneur. The lake floods a collapsed area over the salt, where the salt dissolved in circulating groundwater.

A salt mine operated until November 20, 1980, when it came to a catastrophically abrupt end. An oil rig operating offshore in the shallow waters of Lake Peigneur drilled into the mine workings. The lake quickly drained into the mine, the overburden collapsed, and the drill rig was pulled down. Saturated mud flowed across the lake bottom, while firm clay along the southeast side of the lake slipped in a series of curving landslides. Within a few days, the lake returned to its former level, having filled all the empty space in the cavern and mine. You can still see the landslide area on the shoreline near Live Oak Gardens, as well as inundated and rotated power poles and houses.

This event shows that subsidence above salt domes can happen suddenly as well as gradually. Both types of subsidence have probably occurred naturally several times in the recent geologic past at Jefferson Island.

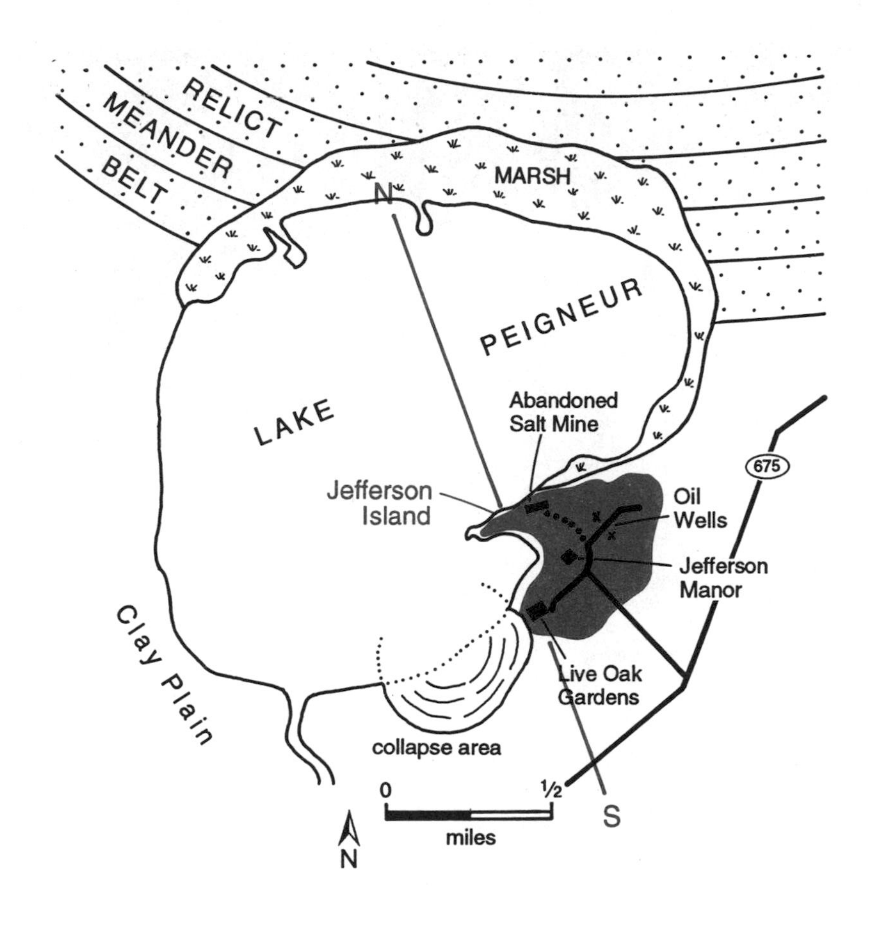

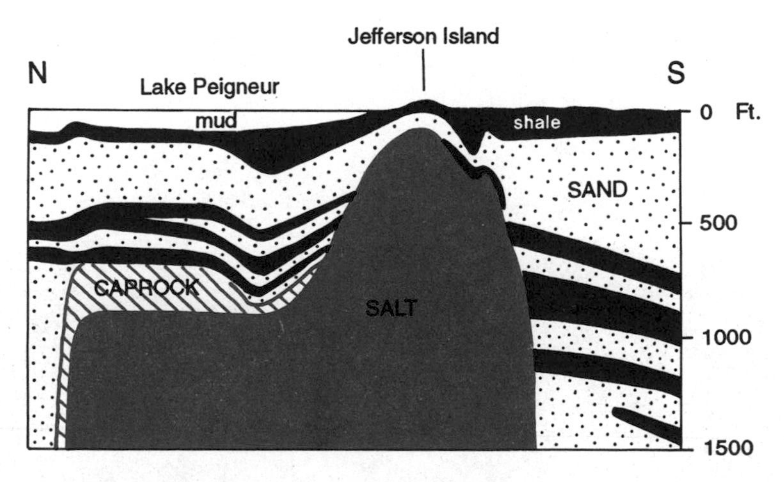

Map and cross section of Jefferson Island. —Autin, 1984

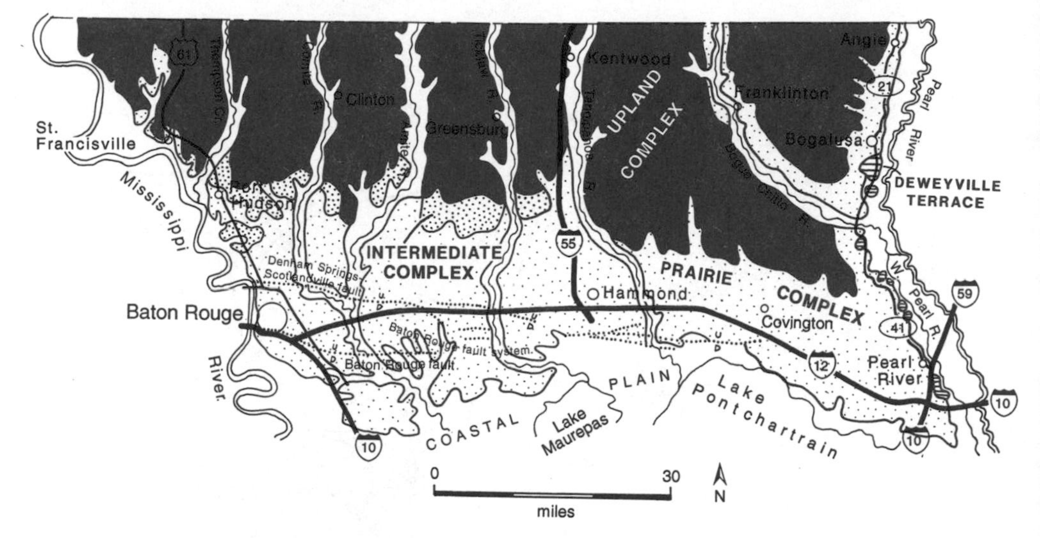

Geologic map of the Florida Parishes in eastern Louisiana.

North-south profile showing the geology of the Florida Parishes (along I-55 from the Louisiana-Mississippi border through New Orleans to the end of the modern Mississippi delta).

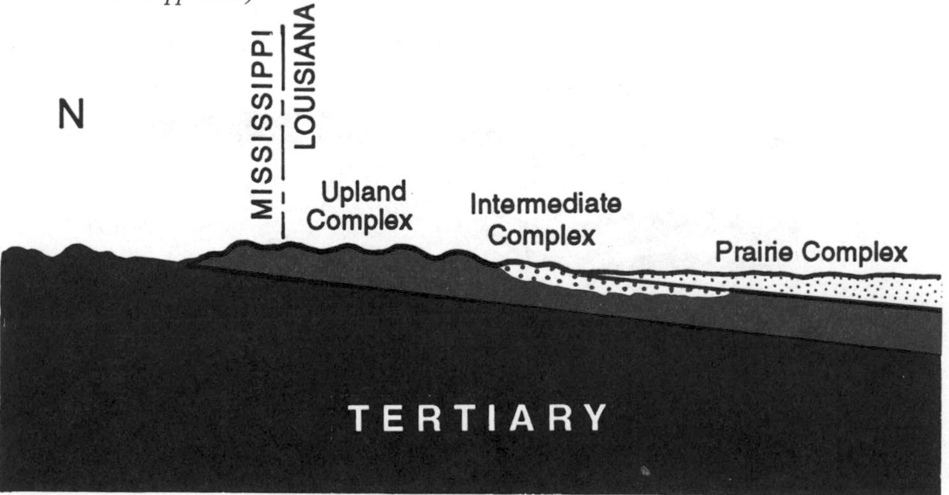

Eastern Louisiana

Red Hills Above Pontchartrain

For those who believe Louisiana is all marshes and swamps, a drive through the Florida Parishes will reveal a different picture. Just a few miles north of the wetlands around New Orleans and Lake Pontchartrain, surprisingly colorful red hills and contrasting dark green pines are tucked away in this eastern corner of Louisiana, the Florida Parishes.

This area between the Pearl and Mississippi Rivers on the east and west, and south of the Mississippi state line to Lakes Pontchartrain and Maurepas, was once the western end of the Florida panhandle. That was when Spain (1565–1763, 1783–1810) and Britain (1763–1783) still claimed Florida. The Florida Parishes were the independent Republic of West Florida for 74 days in 1810, then joined the United States.

An observant traveler in the colorful Red Hills above Lake Pontchartrain can see all the geologic elements of the Florida Parishes. The sedimentary deposits include red Citronelle formation, the Upland,

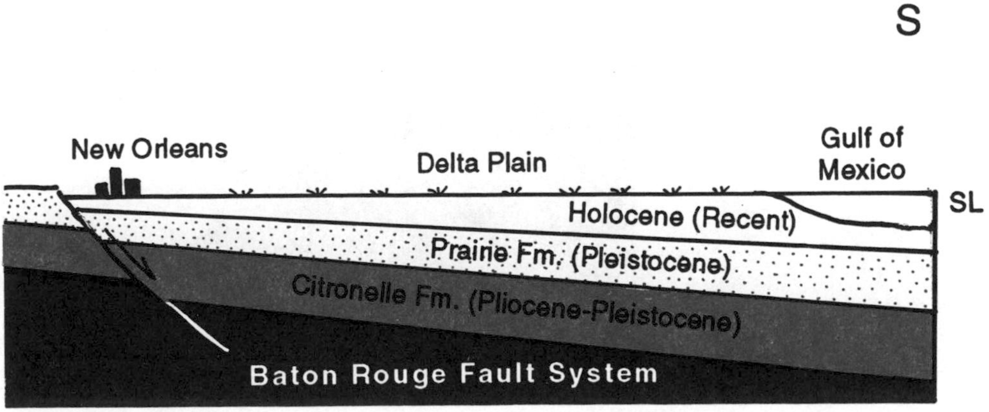

Intermediate, and Prairie Complexes, and a cap of windblown loess. The Deweyville terrace is conspicuous along the Pearl River. The Baton Rouge fault system cuts across them; modern streams and rivers have dissected all of them, the older ones the most.

The Upland Complex

The band of reddish sediment that sweeps across the northern half of the Florida Parishes runs roughly parallel to the coastline and stands high, by Gulf Coast standards, at 150 to 360 feet above sea level. These colorful sedimentary deposits of sand, silt, clay, and gravel are the Citronelle formation, laid down as a broad alluvial apron by coalescing braided streams. Most geologists believe this happened during Pliocene time, just before the great ice ages began. The heavy minerals in the Citronelle formation suggest that the Appalachian Mountains were the likely source for these sediments. Similar deposits in central Louisiana came from the Ouachita Mountains.

In some areas outside of Louisiana, the Citronelle formation is called the Lafayette formation. Under these two names, it caps ridges all over the Southeast, from southern Missouri south to the Gulf of Mexico, and from central Oklahoma east to the south Atlantic coastal plain. In the modern world, such alluvial aprons form only in deserts, where streams do not have enough flow to carry their burden of sediments to the ocean. If observation of the modern world is a reliable guide to conditions during Tertiary time, then the Citronelle formation was laid down on a desert alluvial plain.

When was the Southeast a desert? Although most geologists assign a Pliocene age to the Citronelle formation, some believe it is much older, middle Miocene. All agree that its age falls somewhere in the range between 2 and 17 million years.

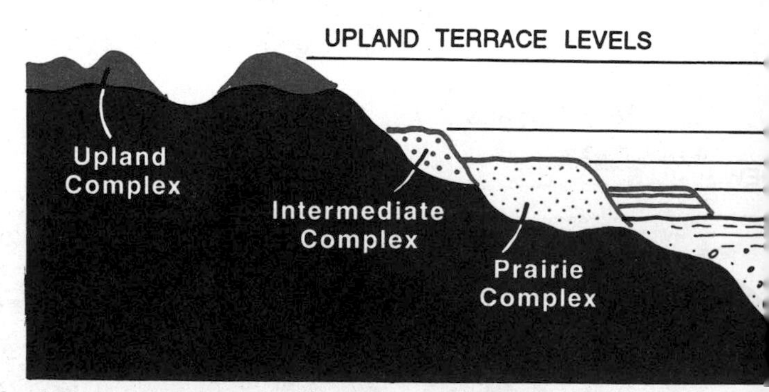

General profile for the lower Mississippi River valley and many smaller adjacent valleys, such as the Pearl River, showing deposits and terraces which reflect different levels of erosion and valley fill during Quaternary and Holocene time. —Autin et al., 1991

On the south Atlantic coastal plain, the Citronelle formation comes to its seaward limit exactly where the middle Miocene formations deposited in seawater come to their landward limit. That relationship strongly suggests a Miocene age. The red soil so deeply developed on the Citronelle also suggests that it is much older than the sediments of the Intermediate and Prairie Complexes, but parts of the Citronelle formation may be the same as the Upland Complex of Pleistocene to Pliocene age.

Whatever the age of the Citronelle formation, it is very deeply eroded. Valleys 200 feet deep between uneroded ridges are common in many places. Earlier geologists thought they recognized terrace levels within this terrain and believed the terraces could be related to Pleistocene warm and cold periods when sea level stood higher or lower than now. Sea level was lower during glacial periods, and streams eroded their beds; sea level was higher during interglacial periods, and streams tended to deposit sediment in their valleys. So the Citronelle formation may be associated with a group of high terraces, but it is hard to be sure how many exist. So the much simpler term Upland Complex is more useful in discussing the elevated terrain of the northern Florida Parishes. The Upland Complex combines the Citronelle formation, river deposits of glacial and nonglacial origins, the upland topography, and the higher and older terraces cut into these deposits, in a unified term, mapped across the state.

The Prairie Complex

A second wide band of lower and younger deposits and terraces lies south of the Upland Complex; it is called the Prairie Complex. This terrain is much less dissected than the Upland Complex, stands 50 to 150 feet above sea level, and is composed of light gray and tan silt, clay, and sand deposited by meandering streams on river floodplains and deltas. The

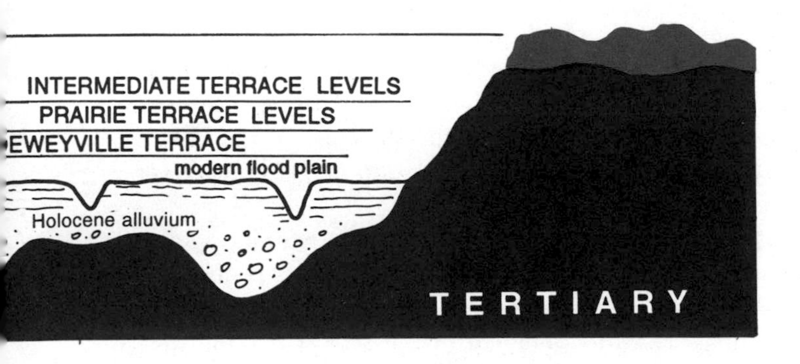

Prairie Complex is made of late Pleistocene deposits, dated by convincing radiocarbon measurements at about 25,000 to 30,000 years.

The Intermediate Complex

A narrow and discontinuous band of terraces and deposits with intermediate sediment characteristics, age, and intermediate elevations lies between the Upland and Prairie Complexes. Geologists map it as the Intermediate Complex.

The Deweyville Terrace

A terrace level between the Prairie Complex and the modern floodplain, the Deweyville terrace, exists in the Pearl River valley. It trends parallel to the Pearl River, not to the coast, as do the Upland and Prairie Complexes.

Loess

Changes between wet and dry climates affect the plant cover, which controls runoff and erosion. Climate and the Mississippi River contributed to deposition of the windblown silt, loess, that blankets the Upland and Prairie Complexes in the Florida Parishes. The loess is tan silt that probably blew out of the Mississippi River valley where abundant dry silt was readily available during dry periods, when the plant cover was sparse. It is thickest, as much as 35 feet, near the Mississippi River, and thins eastward across the Florida Parishes. One layer, the Peoria loess, covered the western half of both the Upland and Prairie Complexes about 20,000 years ago. It is easy to see along roads north of Baton Rouge. An older loess covered only the Upland Complex about 75,000 to 95,000 years ago. It spreads eastward to the Pearl River, but is thin and hard to distinguish from the underlying sediments.

The Moving Land

The Wiggins arch is a broad fold in the rocks north of the Florida Parishes. Detailed elevation surveys over many years show that it is rising at a rate of about one-fifth of an inch per year. That may not seem like much, but if you multiply it by 21,600 years, you get 360 feet, which is the elevation of the northwestern corner of the Florida Parishes. This means all of that elevation could be the result of the slow rise of the Wiggins arch from only late Pleistocene time to the present, or in about the past 20,000 years.

Other surveys reveal that the modern coastal plain is sinking relative to the Florida Parishes. The Baton Rouge fault system is probably the hinge between the rising land in the Wiggins arch and the sinking coast. These movements may account for much of the seaward tilting of the

The red stick (baton rouge *in French*) *was a boundry marker of the Natchez Indians, from which the capital of Louisiana was named.*

terraces. Some geologists have pointed out that the Baton Rouge fault system defines the boundary between terrace levels of the Prairie Complex and the coastal plain in some places. Evidently, the boundary is not a simple matter of sediment deposition, but may also depend on movement along the fault.

Interstate 12
Baton Rouge—Hammond—Slidell
85 miles

This segment of interstate highway skirts the north sides of Lake Pontchartrain and Lake Maurepas. The road is on the Prairie Complex, along the southern edge of the uplands of the Florida Parishes. The elevation is about 50 feet above sea level and the terrain is flat to mildly undulating. These flatwoods of pines and deciduous trees prefer the sandier soils of the Prairie Complex. The Prairie Complex is flat, but it differs noticeably from the lowland swamp around the lakes just a few miles to the south.

In the eastern half of the route, between Slidell and Hammond, the highway crosses the Bogue Falaya, Tchefuncte, and Tangipahoa Rivers, which drain into Lake Pontchartrain from the Upland Complex in the northern Florida Parishes. The western part of the route, between Hammond and Baton Rouge, crosses the Amite, Tickfaw, and Natalbany Rivers, which drain into Lake Maurepas. Many of the riverbank crossings reveal glimpses of tan to gray sand, silt, and clay typical of the sediments of the Prairie Complex.

A quick look into the Amite River nine miles east of the junction with Interstate 12 provides a nice view of sandy point bars. The Amite River has been the subject of much geologic research on how streams respond to changes in sea level, delta formation, climate changes, and subsidence of the land.

Loess

Windblown silt, loess, covers the Prairie Complex in the area between Denham Springs and Baton Rouge. It is thickest, about 35 feet, near the Mississippi River, and thins eastward, which suggests that it probably blew off the floodplain. The age of the main layer, the Peoria loess, is about 20,000 years, near the end of the last ice age. You see very little loess along the road, but you can see it in roadcuts north of Baton Rouge.

Faults

Interstate 12 is north of the Baton Rouge fault system near Hammond, crosses the Denham Springs–Scotlandville fault just east of Livingston, and is near the Baton Rouge fault at the junction with Interstate 10 in Baton Rouge. The Denham Springs–Scotlandville fault apparently gov-

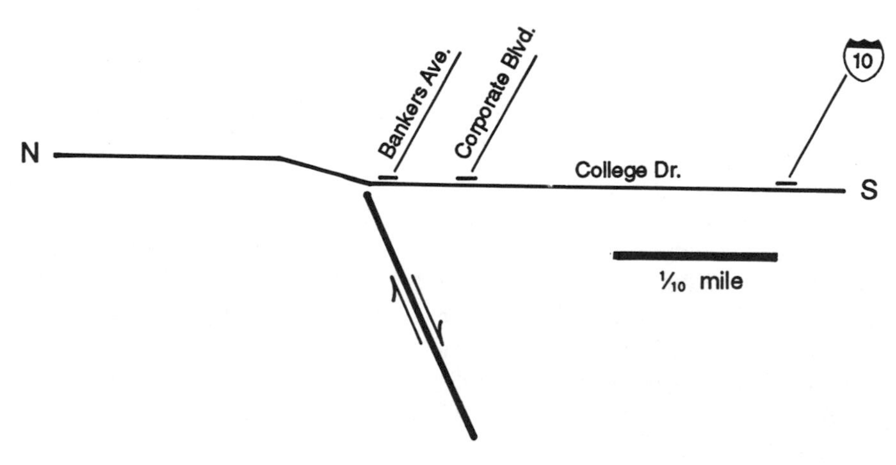

Profile shows surface expression of Baton Rouge fault along College Drive north of I-10.

erns the location of Denham Springs, the mineral springs that William Denham discovered in 1827. Many excellent artesian wells produce from the Prairie Complex in the Florida Parishes. The water enters the ground in recharge areas in southern Mississippi that are 400 to 500 feet above sea level, which provides the hydraulic head that makes the wells flow.

The offset along the Baton Rouge fault system is generally hard to see, but you can get a good view of the Baton Rouge fault near the highway in Baton Rouge. Exit Interstate 10 a mile west of the I-12 junction at College Drive and drive two-tenths of a mile north on College Drive. The abrupt 20-foot rise in College Drive just north of Bankers Avenue is the surface offset of the Baton Rouge fault. Only three miles south, at University Field, the fault is 7,000 feet below the surface. It dips south at an angle of about 70 degrees, which flattens to about 50 degrees or less at 10,000 feet below the surface. The offset at that depth is hundreds of feet.

Interstate 55

Hammond—Kentwood Quarry—Mississippi

38 miles

This segment of Interstate 55 north of Interstate 12 crosses the two main upland aspects of the Florida Parishes—the Prairie Complex and the Upland Complex with its red Citronelle formation. A few miles north of the swamps and marshes around New Orleans and Lake Pontchartrain, you'll enter pine forests and rolling topography nearly 300 feet above sea level. Brilliant red soils accent the green of the pine trees.

Old Dunes

The southbound rest area north of Hammond is on sand dunes that rest on the Prairie Complex. These dunes are Pleistocene in age and appear in several trends closely associated with river meander loops. They were likely built by winds blowing sand out of dry river deposits when the climate was much drier and cooler than now.

Prairie Complex

Between Interstate 12 and the Amite exit, Interstate 55 follows the drainage divide between the Natalbany River to the west and the larger Tangipahoa River to the east. Route 12 is on the Prairie Complex; the mildly undulating landscape averages about 50 feet above sea level. Pine and deciduous trees are commonly mixed in these flatwoods, and swamps exist here and there in low areas around streams. Watch for rare exposures of light tan clay, silt, and sand typical of the Prairie Complex.

Citronelle formation in Kentwood Quarry.

Upland Complex

The road rises to the Upland Complex about a mile north of the Amite exit, and the terrain becomes very hilly. From here to the Mississippi border, the elevation increases steadily to about 250 feet above sea level at Kentwood, and the hills become more rugged. The deeper erosion reflects the greater age of the Upland Complex. Bright red, maroon, and orange soils characteristic of the Citronelle formation contrast the dark green pines in these southern pine hills.

The highway builders desecrated most of the roadcuts along I-55 with grass, so you see very little of the red and orange Citronelle formation. Any side road where the roadcuts were left in their natural condition provides better exposures.

Kentwood Quarry

The Kentwood quarry provides outstanding exposures of the Citronelle formation. Exit east off Interstate 55 onto Louisiana 38 to Kentwood. Turn north on 13th Street; go three blocks, and you will see the quarry to the left. If you do not stop at the quarry, you can see it as you drive by on the freeway; look east just north of the Kentwood exit.

The Citronelle formation typically comes in layers of sand and gravel in shades of red and orange. A film of red iron oxide, the iron minerals hematite and limonite, coats the sand grains and pebbles. Iron oxide is an excellent pigment, so a thin coat on the grain surfaces gives them a bright color. The iron oxide comes from the weathering of iron minerals.

Vertical faces in the upper quarry display steeply angled layers. Some are straight and parallel to each other; others are curved. These are crossbeds, minor layers within the larger layers, and at an angle to them. Fast and turbulent water moved this sand in a stream, shaping it into ridges and waves on the streambed. The crossbeds are the internal structures of these forms. Look also for irregular surfaces beneath the crossbeds, and see that the base of the crossbed sets ends abruptly at them. These are scour surfaces, where streams cut channels into the sediment, then filled them. Purple and white clay pebbles of various sizes scattered in the sand were ripped up from older clay beds by the scouring action of the stream.

Look for dark reddish soil zones with white vertical tubes in them. They are bleached zones around ancient plant roots that grew in the soil. The carbon in the roots bleaches the red iron oxide. The occasional streaks of black carbon in the centers of the tubes are remnants of the original roots.

Between Kentwood and the Mississippi border, I-55 continues across rolling pine hills and red soils of the Upland Complex.

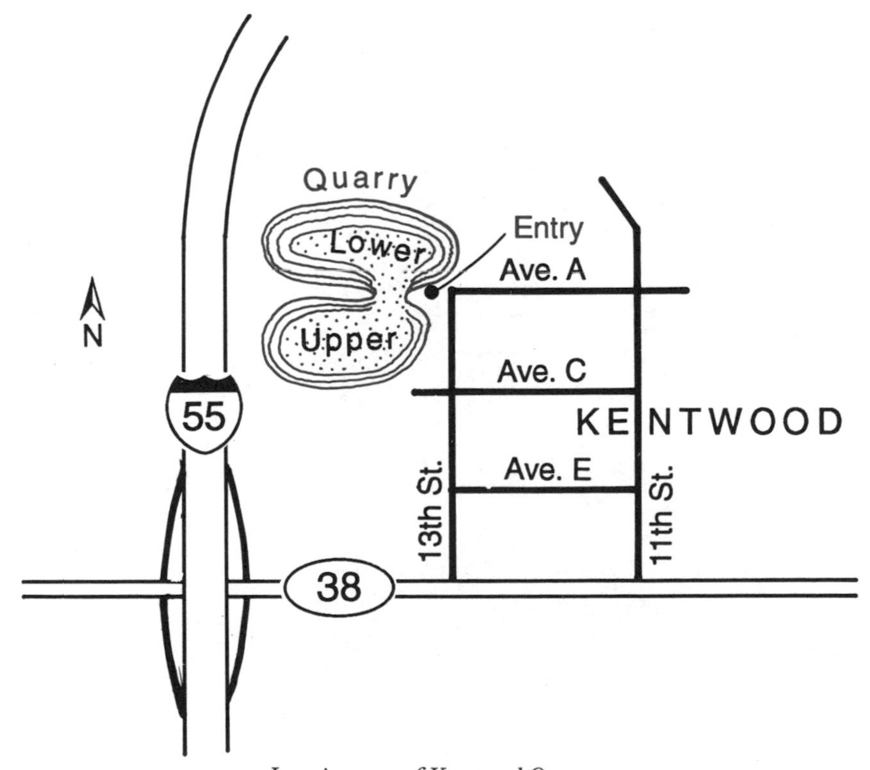

Location map of Kentwood Quarry.

1-*Crossbedding,* 2-*ripples, and* 3-*scour surface in red sand, Kentwood Quarry. Quarter at top center for scale.*

White, bleached root traces in red sand, Citronelle formation in Kentwood Quarry.

Interstate 59
Mississippi—Slidell
15 miles

The Pearl River

Interstate 59 enters Louisiana in the Pearl River floodplain. Within a mile of the border you can see a large sand mining operation south of the road. The sand is extremely white, probably because acid swamp waters dissolved the iron stain off the quartz grains. Backswamp and river channels dominate the scenery. Six miles from the state line, I-59 arches over the Pearl River on a high bridge that provides a good view of the river.

The Pearl River heads in Mississippi and flows south below Jackson to define the eastern border of Louisiana. It is one of the larger streams east of the Mississippi River that drains a watershed entirely on the young rocks of the coastal plain. The east, middle, and west channels of the Pearl River flow along the Louisiana border in tortuous but essentially parallel paths. They drain into Lake Borgne, where the deposits of clay are beginning to build a bay-head delta over older deposits of the St. Bernard delta of the Mississippi River. Minerals in the Pearl River sand, eroded from regional Tertiary and Pleistocene sediments, originally came from the Appalachian Mountains.

The exit for the town of Pearl River is at the western end of the bridge. The vegetation and topography there change abruptly from tangled swamp to firm ground covered with pine trees. The highway is on a prong of the Prairie Complex that arcs around the eastern end of Lake Pontchartrain. The road continues on the Prairie surface to Slidell.

US 61
Baton Rouge—St. Francisville—Mississippi
36 miles

Highway 61 north of Baton Rouge is a scenic drive into the colorful Tunica Hills. Red sediments and tan loess lie beneath rolling hills in this northwestern corner of the Florida Parishes. Some of the hills rise to 350 feet and many stream valleys are 200 feet deep. This road crosses all four sedimentary elements of the Florida Parishes: the Upland Complex with its Citronelle formation, the Intermediate Complex, the Prairie Complex, and deposits of loess.

Through St. Francisville, Wakefield, and Laurel Hill to the Mississippi border, US 61 climbs through the Tunica Hills of West Feliciana Parish. The Citronelle formation is the bedrock, and the uplands display the characteristic red hues and hilliness of Upland Complex landscapes.

The more gravelly parts of the Citronelle formation cap the hills. Some geologists suggest that the hills are where they are because the gravel differentially protected them from erosion. Gravel, because it is highly permeable to water, is extremely difficult to erode—provided it is above the reach of a stream or waves.

Loess mantles the Upland Complex—on some hilltops as deep as 35 feet, but 3 feet is more typical. You can see an excellent exposure of loess at the top of a bluff cut in natural levee deposits at the western end of Ferdinand Street in St. Francisville, where it goes down to the river. The vertical face, tan color, and fine but gritty texture of the loess are diagnostic. If you roll a bit of it between your fingers it feels like fine powder, but it grits between your teeth.

Six miles of highway south of St. Francisville cross the Upland Complex, with the Citronelle formation underfoot. Six miles south of St. Francisville, the road crosses Thompson Creek, the border between West and East Feliciana Parishes. If the water is low, look for sandy point bars in the creek.

The highway between Thompson Creek and Baton Rouge is mainly on the Prairie Complex, except for short segments that cross the Intermediate Complex near Irene. Several roadcuts a bit less than a mile south of Thompson Creek expose Prairie Complex sediments with a cap of deeply gullied loess. Loess naturally weathers to nearly 90-degree slopes that will withstand weathering for many decades, but lower angle cuts along highways encourage it to erode into ravines and gullies.

The road is on the Prairie Complex at Port Hudson, where a Civil War battle was fought in 1863. Part of the battlefield is on the Prairie Complex, the rest on a natural levee. Streams cut deep ravines into the loess on the terrace behind the high Prairie Complex bluffs, which made Port Hudson an ideal natural defensive position. Big guns on the high bluffs commanded the great loop of the Mississippi River. Short hiking trails at the battlefield park go through cuts and ravines where you can see alluvial sediments of the Prairie Complex with their cover of loess. The

Classic, sandy point bar on inside bend of Thompson Creek. Photo taken from US 61 bridge.

Erosion rills in Pleistocene Prairie Complex silt. Roadcut just south of Thompson Creek.

Civil War earthworks have not been restored; they are just nicely preserved. The defenders placed these structures to use the topography to best advantage.

Louisiana 21, Louisiana 41
Mississippi—Bogalusa—Interstate 59
54 miles

This easternmost road of the Florida Parishes follows the valley of the Pearl River, which is not visible from the road. Most of the road is on the relatively flat surface of the Prairie Complex. Occasional exposures in shallow roadcuts and stream banks reveal tan sandy sediments and soils. Pine woods dominate the countryside, and supply the timber mills. Side roads that head west climb onto the red and dissected hills of the older Upland Complex.

About two miles south of Bogalusa, the road dips subtly onto the slightly lower surface of the Deweyville terrace. It is lower and younger than the Prairie Complex, but higher and older than the floodplain of the Pearl River. The Deweyville terrace is a late Pleistocene deposit laid down between 30,000 and 17,000 years ago, during the last ice age. The Pearl River was much larger then, and sea level was much lower. Gray silt, clay, and sand appear along the road. Farther south, the road drops onto the swampy terrain of the modern floodplain.

A few miles farther south, between Sun and Bush, the road crosses the valley of the Bogue Chitto River. This stream carries a lot of sand and some gravel, which it erodes from the Citronelle formation in the Upland Complex country to the northwest. Sand and gravel are in great demand on the Gulf Coast for construction and road aggregates.

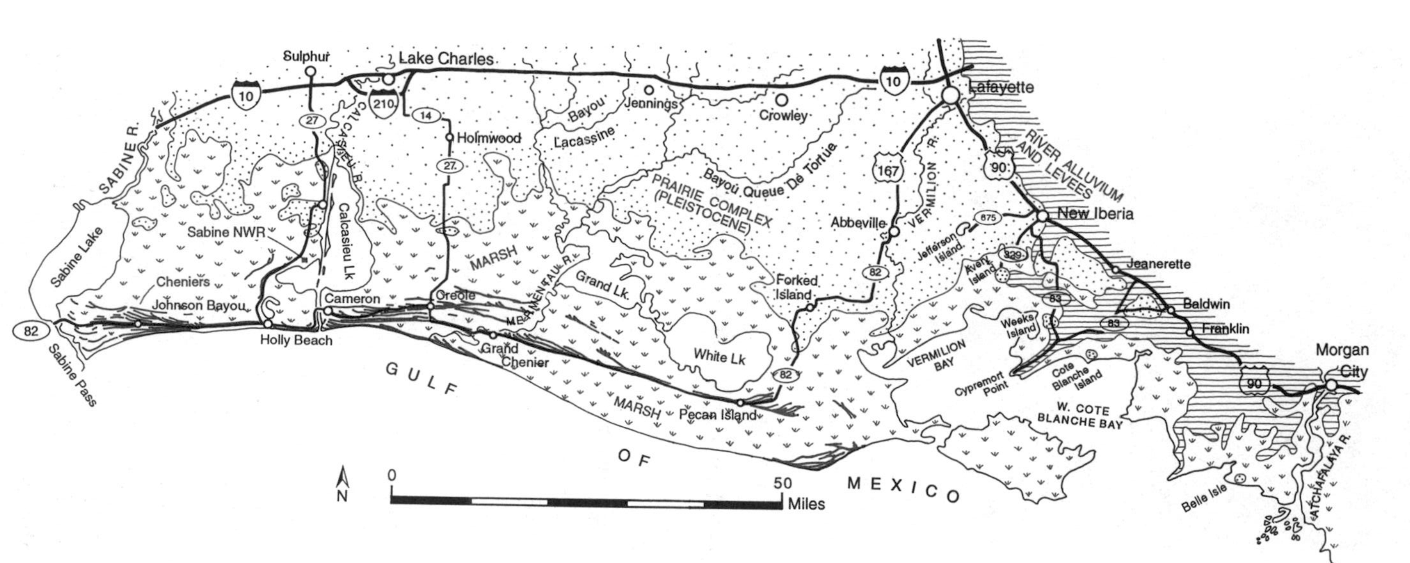

Geologic map of southwestern Louisiana.

Southwestern Louisiana

Marsh to the Sea

Southwestern Louisiana is a land of almost minute topographic changes, but important environmental distinctions. Most larger towns, much of the commerce, and many of the major highways are on the Prairie Complex, not much above sea level, but enough to avoid most of the floods and hurricanes that surge across the marshes bordering the Gulf. The Pleistocene Prairie Complex and the marshes of the modern coastal plain are no more than about 50 feet apart in elevation, but they define the two landscape styles of southwestern Louisiana.

Interstate 10 rolls in a straight line across the high, dry Prairie Complex between Lafayette and Lake Charles, obliquely crossing drainages making their way to the coast.

The Chenier Plain

Roads leading south from Interstate 10 drop off the Prairie Complex onto the chenier plain, a vast expanse of fresh and salt marshes that extend in all directions from the road to the horizon. Sturdy grasses nod in the breeze, beckoning hordes of waterfowl to roost and feed in the rich organic muck. Lakes and long ribbons of elevated sand ridges topped by magnificent live oak trees punctuate the marshes. They seem out of place in this marsh-to-sky milieu, but the ridges, the cheniers, give the plain its name. The French word *chene* means oak, hence a chenier is a ridge forested with oaks.

The chenier plain differs markedly from the delta plain to the east; both are distinct parts of the modern coastal plain. Maps that show the complicated lacework pattern of coastal bays, barriers, and eroded headlands of the delta plain look very different from those that show the smoothly curved shoreline of the chenier plain. Evidently, different geologic processes affect the two plains.

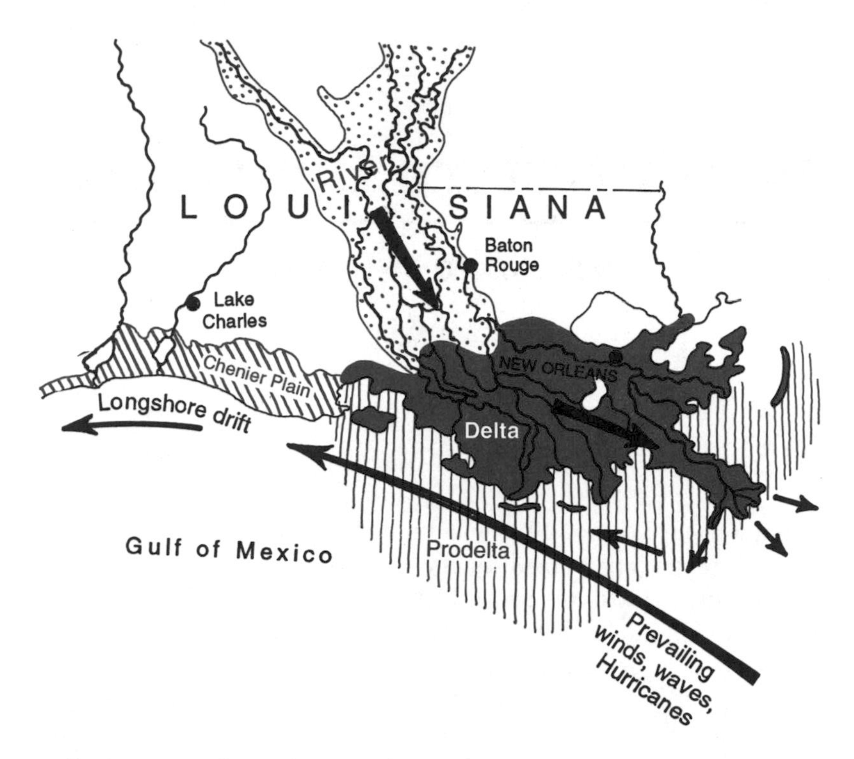

Environments where sediment is deposited and patterns of sediment dispersal in Gulf waters. Predominant winds from the southeast move sediment northwestward from the Birdfoot delta. —Scott, 1969

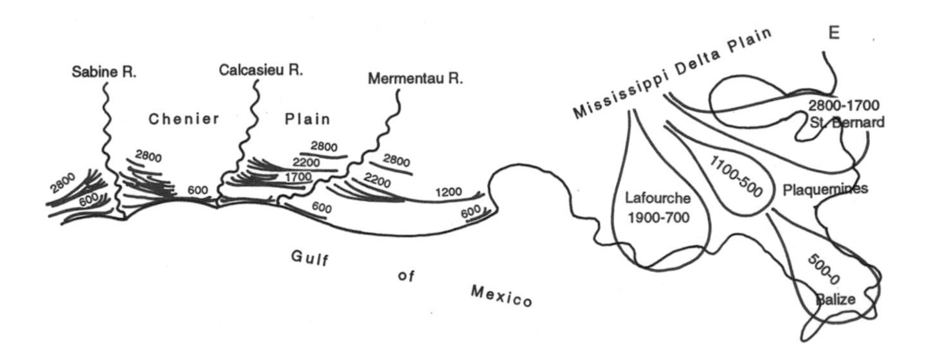

Position and dates of Mississippi delta lobes compared with dates of cheniers. Sketch is not to scale. Delta lobes older than 2,800 years are not shown. Note concentration of cheniers at mouths of rivers.

The delta plain forms where the Mississippi River and its distributaries pour a relentless torrent of muddy sediment into the Gulf of Mexico. No major river now feeds sediment into the chenier plain, though it stands on a platform of older deltas. Longshore currents that sweep westward from the Mississippi delta import the mud and sand that are building the chenier plain seaward.

The chenier plain is a flat expanse of low marsh and mudflats, about 120 miles long by 12 miles wide. Ribboning through the mudflats are chenier ridges of sand and shell hash, grouped in bundles that parallel the coast. Radiocarbon dates reveal that the oldest ridges, about 2,800 years old, are farthest inland, and that they become progressively younger toward the present shoreline. The chenier plain formed in the past few thousand years, after sea level rose to about its present level.

After they had obtained radiocarbon dates on the cheniers, geologists thought they could see a rough correlation with the dates of the various delta lobes of the Mississippi River. They proposed that when the Mississippi River was building a delta in the western part of its valley near the chenier plain, plenty of mud was available for longshore currents to sweep westward, building the plain seaward. When the Mississippi River shifted eastward, the nearby mud supply was cut off and the chenier plain stopped growing. Instead the waves eroded the chenier coast, reworking sediment, and depositing the eroded coarser grains of sand and shell hash as narrow beaches against the muddy marsh. When the delta shifted westward again, the process repeated, and muddy marsh began to grow seaward once more, enriched by a new, nearby mud supply. Each sandy chenier ridge then records a period of sediment starvation and wave erosion, each muddy marsh between ridges a period of sediment deposition and growth. It was a nice theory.

More recent research produced more data, which suggest that the story may be a little more complicated. It is now clear that the chenier plain contains more sandy ridges than the delta has lobes. And some chenier ridge dates fall right in the middle of the range of dates for phases of westerly deltas, when mud, not sand, should accumulate. So it now seems that no precise correlation exists between times of delta growth and growth of the chenier plain.

Fluctuations of mud and sand supply to the chenier plain may reflect periods of high and low floods in the Mississippi drainage, as well as simple delta shifts. This would provide more changes of mud and sand supply than delta shifts could. Bundles of chenier ridges appear to concentrate around the mouths of the Sabine, Calcasieu, and Mermentau Rivers, which now cross the chenier plain. Perhaps these rivers may be the source of sand to build cheniers, rather than waves reworking the marshes.

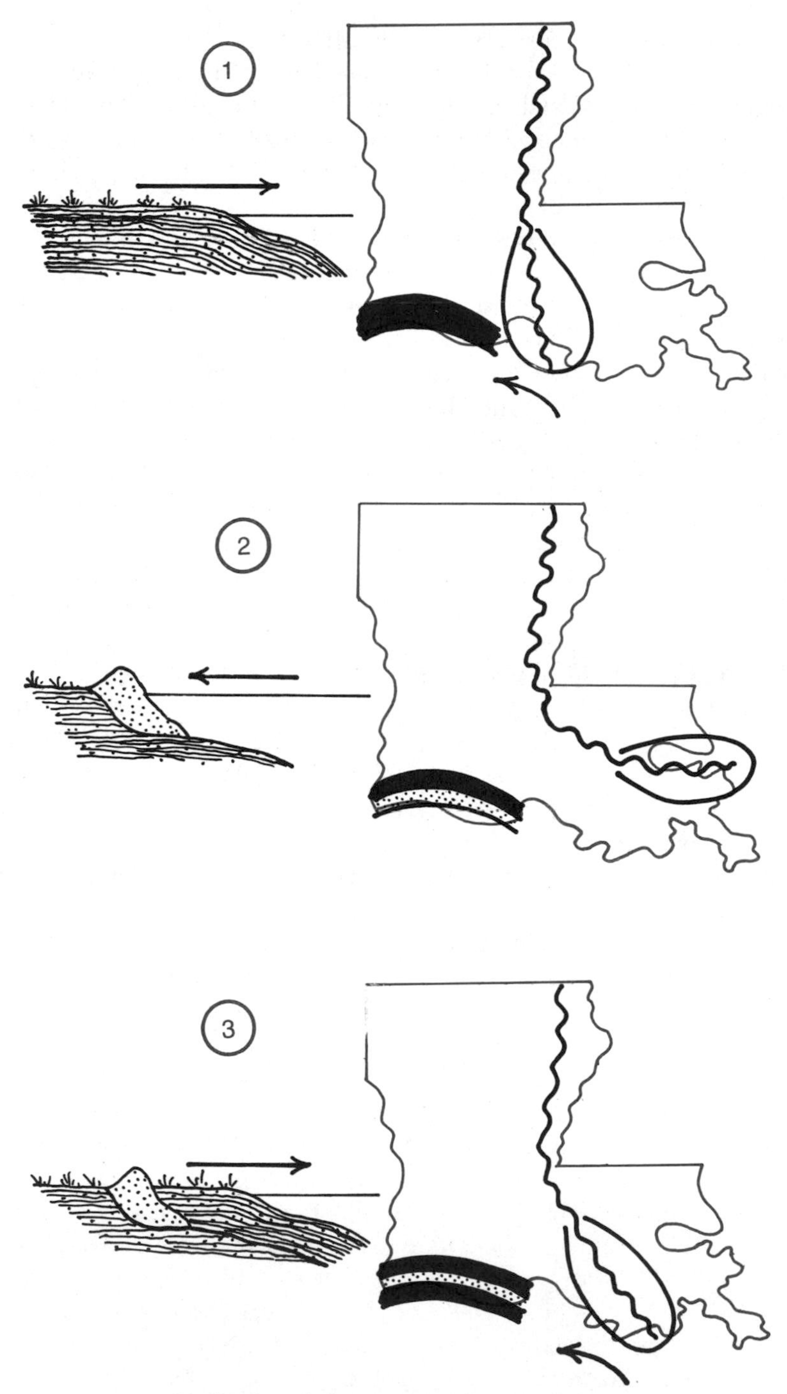

Sand and mud alternations of the chenier plain.

1 Longshore currents add mud from nearby delta; chenier plain grows seaward.

2 Delta shifts away; mud supply is cut off; chenier plain is stable or cut back by waves; sand ridge (chenier) builds at shoreline.

3 Delta shifts back; again mud is moved along shore and added to chenier plain.

The marshes of the chenier plain are marvelously dynamic places that combine the mineral richness of the underlying muds with stored energy of the abundant grasses and water plants. Larval and juvenile stages of marine animals from crabs and shrimp to fish abound. Waterfowl feed and nest amid a bountiful supply of food, where water and a good view to the horizon protect them from most predators. About 6 million migratory birds winter in these marshes; others stop before crossing the Gulf of Mexico on their way to the Yucatan and places farther south.

Interstate 10

Lafayette—Jennings—Lake Charles—Sulphur—Texas

105 miles

Except where it crosses the Sabine River, the highway traverses the Pleistocene Prairie Complex along the entire stretch between Lafayette and Texas. The Prairie Complex was the coastal plain during the last ice age, and its sediments are similar to those on the modern coastal plain: river, delta, floodplain, and backswamp deposits. If you look closely at riverbanks along the freeway, you will see the gray to tan mud, silt, and fine sand typical of the Prairie Complex. The average elevation along the way is about 40 to 50 feet, most easily discernible when you drive off the terrace onto a lower floodplain of a modern river. The road is quite straight and the landscape reasonably flat, except for dips into occasional stream valleys.

Loess

A yellowish deposit of windblown silt, or loess, covers the Prairie Complex between Lafayette and Rayne. It probably was blown off the floodplain of the Mississippi River during Pleistocene time. Geologists who have mapped the traces of Red River channels across this portion of the Prairie Complex have shown that they track southwesterly and have much smaller meander loops than the old Mississippi channels on the Prairie Complex southwest of Lafayette.

Sugarcane grows on the loess that covers the Prairie Complex around Lafayette, whereas rice and soybeans are the common crop west of Rayne. Many rice ponds west of Lafayette also produce crawfish, which are harvested before the rice is planted. These delicious little beasts are often featured in exquisite Cajun cuisine, famous in Louisiana around Lafayette, and Creole cooking, centered in New Orleans.

Timbered cypress stumps in Bayou des Cannes.

Between Crowley and Jennings, the highway crosses Bayou Nezpique, Bayou des Cannes, and Bayou Plaquemine. The three bayous are tributaries of the Mermentau River, which empties southward into Lake Arthur and Grand Lake, both on the modern coastal plain.

Oil and Gas

The other big cash crop along I-10 is oil and gas; watch for the nodding pumps that produce the oil and the gas well heads with all their valves and gauges. The Louisiana Oil and Gas Park at Jennings borders the north side of I-10. Turn off the highway at Exit 64, Louisiana 26. A wooden derrick commemorates Louisiana's first oil well, the Jennings Oil Company #1, which Jules Clement drilled near Evangeline, a few miles northeast of the park, on September 21, 1901. It closely followed the first great gusher in Texas, at Spindletop in January 1901.

Calcasieu River

Interstate 10 crosses the Calcasieu River on a high bridge at the western end of Lake Charles. It provides good views of cypress swamps, the meandering river, and Lake Charles. Sand was hauled in to make the beach on the south side of the freeway.

The Calcasieu ship channel, which essentially follows the course of the Calcasieu River, made Lake Charles the third largest port in Louisiana. The steel forest of refinery towers west of the bridge makes a blazing light show at night.

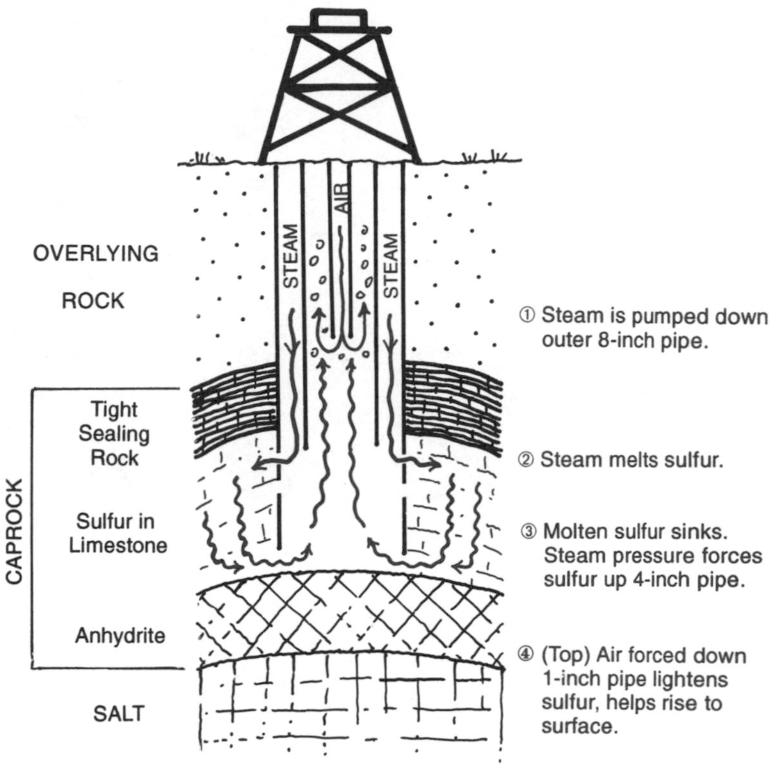

The Frasch process uses superheated steam to recover sulphur in liquid form, which can be pumped to the surface.

Louisiana Oil and Gas Park at Jennings.

Landslide-slump on I-10 overpass west of Sulphur.

Sulphur

Ten miles west of Lake Charles is Sulphur, named for its chief product. This is where sulphur was first produced through steam-injection wells. Herman Frasch injected steam into the caprock of the Calcasieu salt dome at Sulphur in 1894, and pumped molten sulphur out. This broke the back of the Sicilian sulphur monopoly, and established Louisiana as one of the three major producers, along with Texas and Russia. Sulphur is necessary in making sulphuric acid, fertilizers, rubber, paper, and many other products.

Sulphur comes from the caprock of salt domes. An oil drilling rig sinks a hole and sets a one-inch pipe inside a four-inch pipe inside an eight-inch pipe. Steam at 325 degrees is pumped down the outer pipe and into the sulphur, which melts at a temperature of 235 degrees. The molten sulphur sinks because it is heavier than water, and the steam pressure drives it up the inner pipe. Air jetted down the center pipe froths the rising molten sulphur to make it lighter and help drive it to the surface. The molten sulphur is collected in heated tanks and transported in insulated pipes, barges, tank trucks, and ships. Louisiana sulphur attracts many industries.

The Brimstone Museum in Sulphur features exhibits and core samples of salt dome rocks and minerals. Drive one mile north on Louisiana 27 at Sulphur, turn at the Frasch Park Museum sign, and drive another half mile to the museum in the old train station.

Cypress trees and their "knees" in swamp at the visitor information center on the south side of I-10 at the Louisiana-Texas border.

Swamps and Marshes

At the visitor's information center on the south side of I-10, just inside Louisiana, is a boardwalk through a cypress swamp. It provides a wonderful view of the swamp with its cypress trees, cypress knees, maples, ferns, arrowleaf, and animals. The organic productivity of such swamps is very high.

The expressway arches over the Sabine River on a high bridge, where views north and south reveal swampland in sharp contrast to the higher ground of the surrounding Prairie Complex.

The Sabine River

The Sabine River heads in the uplands of East Texas, then turns south to define the border between Texas and Louisiana. Though it now drains a relatively small area, it was once larger and reached much farther inland. Upriver, at Toledo Bend, geologists have found grains of black chromite, a rare chromium mineral; garnet, which comes from igneous and metamorphic rocks; and long crystals of quartz in Miocene sediments of the Sabine River. These minerals almost certainly came from the Ouachita Mountains in Arkansas.

The Red River ran down the Sabine River (dotted shortcut) *in Miocene time. A subtle shift in the Sabine uplift may have tilted drainage of the Red River toward its present course, which joins the Mississippi River.* —Manning, 1990

The Sabine River receives no water from the Ouachita Mountains, but the Red River does. During Miocene time, a few million years ago, the upper Red River and its tributary, the Little River, flowed out of the Ouachita Mountains and down the present course of the Sabine River. Then the lower Red River eroded headward and captured the upper Red River, leaving the Sabine River without a large part of its watershed, and much diminished. No one knows when or why this happened. Some geologists suggest that a small rise in the Sabine uplift may have caused it; others blame shifts in the course of the Mississippi River.

Louisiana 27
Sulphur—Hackberry—Holly Beach
35 miles

This drive offers an opportunity to compare the landscape of the Prairie Complex with that of the lower marshes of the modern coastal plain. The northern third of the road, ten miles south of Sulphur, is on the gently rolling and forested Prairie Complex, a Pleistocene surface about 50 feet above sea level. The land is dry, trees are mostly the common deciduous varieties, and the soils are light tan.

The high bridge over the Intracoastal Waterway ten miles south of I-10 provides a nice vantage point to see the abrupt transition between Prairie Complex and salt marsh. The bridge is on a patch of elevated Prairie Complex, surrounded by marsh. Sand on the Prairie Complex is part of an old beach ridge that records the position of a Pleistocene shoreline. Waterways to the east are part of the Calcasieu River, the Calcasieu ship channel, and the northern end of Calcasieu Lake. Ships using the Intracoastal Waterway can reach Lake Charles or the Gulf of Mexico from this area.

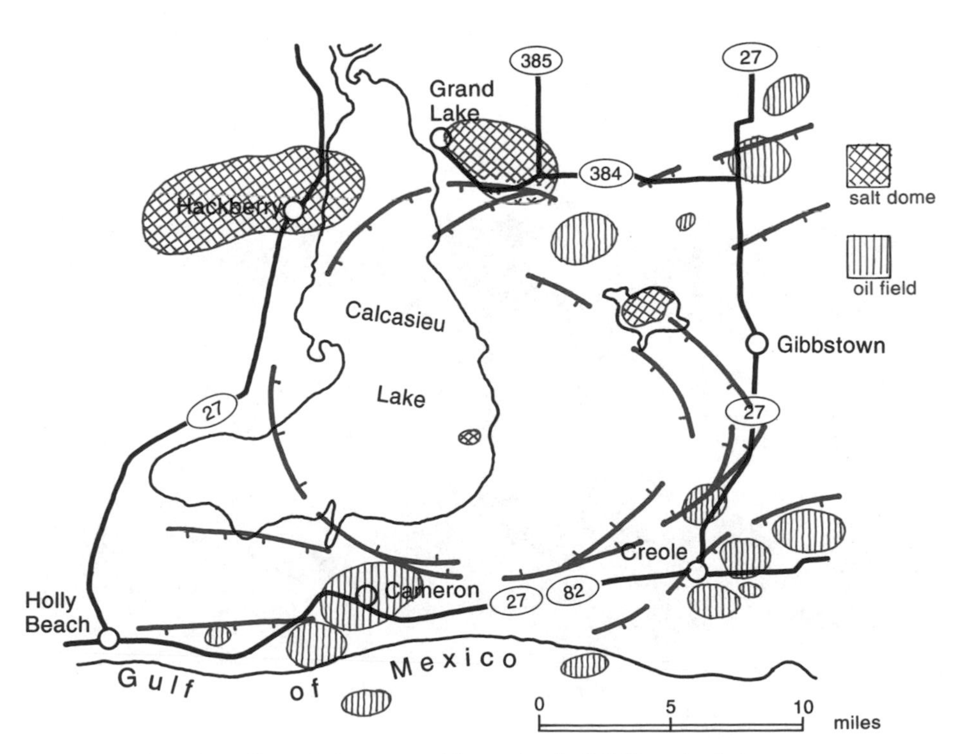

Circular collapse fault system with associated oil fields and salt domes around Calcasieu Lake. —Seglund, 1974

Sabine National Wildlife Refuge visitor center.

Freshwater marsh. Photo taken from boardwalk south of visitor center.

The road south of the bridge follows the Calcasieu ship channel for a couple of miles, at the level of the marsh. The topography east of Louisiana 27 is dredge spoils from the ship channel. Calcasieu Lake is east of the ship channel, but not visible from the road.

Hackberry stands on the high ground of another patch of Prairie Complex. Notice the elevation difference between marsh and prairie at the northern edge of town. The oil wells produce from structures around the Hackberry salt dome, which directly underlies the town and the prairie. Its rise probably contributed to the elevation of the prairie. The geologic map shows that Calcasieu Lake, the Hackberry salt dome, and other salt domes in the vicinity cluster around a large collapse structure.

Louisiana 27 crosses the Prairie Complex for about five miles south of Hackberry, then crosses three miles of treeless salt marsh at an average elevation of about two feet above sea level. It then turns southwest at the Sabine National Wildlife Refuge visitor's center. The refuge was established in 1937 to protect 143,000 acres of fresh and salt marsh, wintering grounds for waterfowl from the central and Mississippi flyways, and an important nursery area for many marine organisms. It offers displays in the visitor's center, and a boardwalk in the marsh four miles to the south.

South for six miles from the boardwalk, freshwater marsh borders the road, which is on a slightly elevated sandy chenier. Open water to the east is West Cove of Calcasieu Lake. Salt marsh lies along the road two miles north of Holly Beach. The town and beach are on a beach ridge that is about 600 years old, according to radiocarbon dates. Cheniers inland of this one are nearly 3,000 years old.

Louisiana 27/14
Lake Charles—Oak Grove
35 miles

Prairie Complex, fresh marsh, and a series of forested chenier ridges are the main geologic features of this roadside tour. From I-210 in eastern Lake Charles, through Holmwood to the Gibbstown bridge 14 miles south of Holmwood, Louisiana 27/14 travels on the Prairie Complex. If you look carefully you can spot the characteristically tan clay soils exposed in ditches between I-210 and Holmwood. Old Pleistocene beach ridges have been mapped on the Prairie Complex about six miles south of Holmwood, where Louisiana 14 makes a sharp jog to the west. From the high Gibbstown bridge you can see fresh marsh stretching southward to the horizon and forested Prairie Complex landscape to the north. The bridge spans the Intracoastal Waterway, which skirts the southern boundary of the Cameron-Prairie National Wildlife Refuge.

Tall cane and wide expanse of marsh, typifying Louisiana's chenier plain.

S MSL beach dunes shaped trees marsh chenier 82 N

Profile along Cameron Parish Road 532 where sandy, oak-covered cheniers alternate with muddy marsh deposits.

Chenier Ridges

South of the Intracoastal Waterway, open fresh marshes of the chenier plain extend to the horizon on either side of the highway. But, five miles south of the bridge where Louisiana 27 turns southwest, a small road heads east toward a line of oak trees in the distance. Why are these trees in the middle of an otherwise wet and treeless marsh? They mark the first of a series of old beach ridges, cheniers.

The sand in the ridges is above the marsh, so the oak trees can grow in the drier soil. This ridge, Little Chenier, marks the position of the beach of 2,800 years ago. A few miles farther on, the town of Creole strings out along another chenier complex where Chenier Perdue, 2,500 years old, and Pumpkin Ridge, 2,200 years old, merge. A few miles farther is Oak Grove, also built on a chenier called Oak Grove Ridge, dated at 1,200 years. The small elevation above the marsh is enough to keep roads, houses, and oak trees above marsh water and occasional storm floods most of the time. People build on the cheniers if they want to live in the chenier plain.

Chenier elevated six to eight feet above marsh, near Sabine Pass causeway. Trees preferably grow on chenier sand.

Louisiana 82
Sabine Pass—Holly Beach—Cameron
36 miles

The Holly Beach area and the drive along Louisiana 82 to the southwestern corner of the state provide one of the best views of a Louisiana chenier. A series of these old beach ridges hugs the eastern bank of the Sabine River in a splayed wedge of sandy ridges, which coalesce eastward into a single main ridge that follows the coastline to Holly Beach. The cheniers stand one to ten feet above the surrounding grassy marsh, enough to give a firm footing to lines of trees that visually identify them from the highway.

The road follows an artificial levee along the two miles east of the Sabine Pass causeway. The houses in Johnson Bayou, about ten miles from Sabine Pass, are built on the chenier north of the road. The line of trees to the south marks another chenier, separated from this one by an expanse of marsh.

Cameron Parish Road 532 leads south from Johnson Bayou to the coast. The road crosses two cheniers and their intervening marshes, then crosses the dunes onto the beach. The form and width of the cheniers are clearly visible, as are the oak trees that grow on them. Salt spray blown in on the sea breeze shaped the oak trees on the chenier nearest the coast.

Wind ripples cover the windward surfaces in the dunes. Look closely to see that the coarse but lightweight fragments of white shells are concentrated on the crests of the ripples, and finer but heavier quartz grains lie in the troughs between ripple crests. The wind sorts particles very carefully.

Holly Beach. Note houses built on stilts.

Offshore oil platform in dry dock at Sabine Pass.

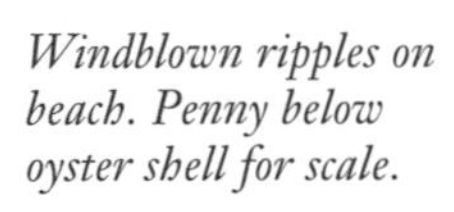

Windblown ripples on beach. Penny below oyster shell for scale.

Five miles east of Johnson Bayou, Louisiana 82 passes a natural gas processing plant, and a few miles farther east the beach community of Constantine Beach stands tall on stilts. They look odd on calm days, but not during heavy storms. Shaped trees tell of relentless onshore winds where Louisiana 82 runs close to the coast east of Constantine Beach. Look south to see offshore oil production platforms on the horizon; most of the wells produce from sandy reservoirs tilted up against the flanks of salt domes.

Look a few miles east of Constantine Beach for riprap blocks of dark gray schist, metamorphic rocks shiny and flaky with mica. They certainly did not come from Louisiana. Shelly concrete boulders and blocks of pale limestone also appear in the riprap. Despite the riprap, waves have eroded gaping patches in the roadway.

Holly Beach is a recreation town, also built on stilts. Flat salt marsh stretches landward behind the Holly Beach chenier on which the town and highway are constructed. It is about 600 years old. Between Holly Beach and Cameron, Louisiana 82 continues to follow the cheniers, which spread out into several, separate ridges, especially noticeable where the road turns north to parallel the Calcasieu ship channel. A ferry takes cars over the ship channel to Cameron. The awful smell you detect as you wait for the ferry comes from the nearby pogy factory. Pogy is a local name for the Atlantic menhaden, a large herring that is by weight Louisiana's largest fish crop. By dollar value, shrimp is the largest catch.

Sand pit in Grand Chenier, east of Oak Grove.

Road gravel composed of white shell fragments.

Louisiana 82, US 167
Cameron—Pecan Island—Abbeville—Lafayette
130 miles

Vast expanses of fresh marshes, large lakes, and long wisps of chenier sands strung across the grassy flats typify this wilderness route across the great chenier plain of southwestern Louisiana.

Along nearly the entire distance between Cameron and Pecan Island, Louisiana 82 follows one chenier or another through the marsh. At the western end of that stretch, between Cameron and Creole, cheniers make a cluster of ridges, and the trees and houses fan out. Halfway between Cameron and Pecan Island, the road turns sharply and crosses cheniers in a roller coaster of ups and downs.

South of Creole, the highway crosses Creole Ridge, a cluster of cheniers that range in age from 2,200 to 2,500 years. Farther south, it crosses an expanse of marsh, and the prominent Oak Grove chenier, which is 1,200 years old, at the town of Oak Grove. You can see that the buildings and trees are on the ridge crest. Watch at the eastern end of Oak Grove for a sand pit north of the road.

Louisiana 82 crosses the Mermentau River, then follows the crest of Grand Chenier, which is 1,200 years old. Multiple cheniers spread northeast of the town of Grand Chenier, just east of the Mermentau River. Rockefeller National Wildlife Refuge lies between the road and the coast between Grand Chenier and Pecan Island. Watch for white road gravel in this stretch, made entirely of broken seashells.

Fifteen miles east of Grand Chenier, the road crosses marsh in an area where the cheniers fade. Farther east, it climbs onto the Pecan Island Chenier, also 1,200 years old. The town of Pecan Island is at the eastern end of the chenier complex, another place where the cheniers fan. In this area, they range in age from 1,200 to 1,700 years.

White Lake, one of the largest bodies of fresh water in Louisiana, is just north of Louisiana 82, but it's not visible from the road. The highway turns north around White Lake and crosses seven miles of marsh.

Eight miles south of Forked Island, the road climbs onto a forested patch of Prairie Complex, which it follows to Lafayette. The high bridge over the Intracoastal Waterway at Forked Island provides an aerial view of a tangle of waterways and small streams that dissect the Prairie Complex.

Very large abandoned meanders of the Pleistocene Mississippi River scar the Prairie Complex between Lafayette and Forked Island. They explain the gently rolling character of the landscape. US 167 follows the Vermilion River between Abbeville and Lafayette, leaving the Atchafalaya basin near Lafayette, then crossing the Prairie Complex in a southwesterly direction to enter the Gulf of Mexico at the western end of Vermilion Bay. The river eroded headward from Vermilion Bay across the Prairie Complex until it cut through to Bayou Teche. It probably became a seasonal distributary when the Mississippi River flowed through the old Teche channel. The Vermilion River now drains the bayou between the Teche levees and the Prairie escarpment.

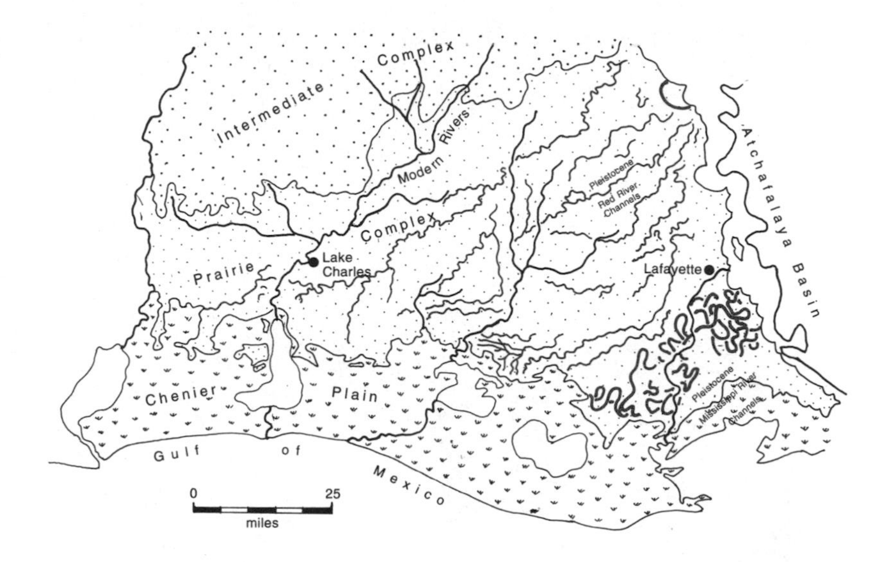

Old channels on the Prairie Complex, showing how the Red River and Mississippi River systems helped build the Pleistocene surface.

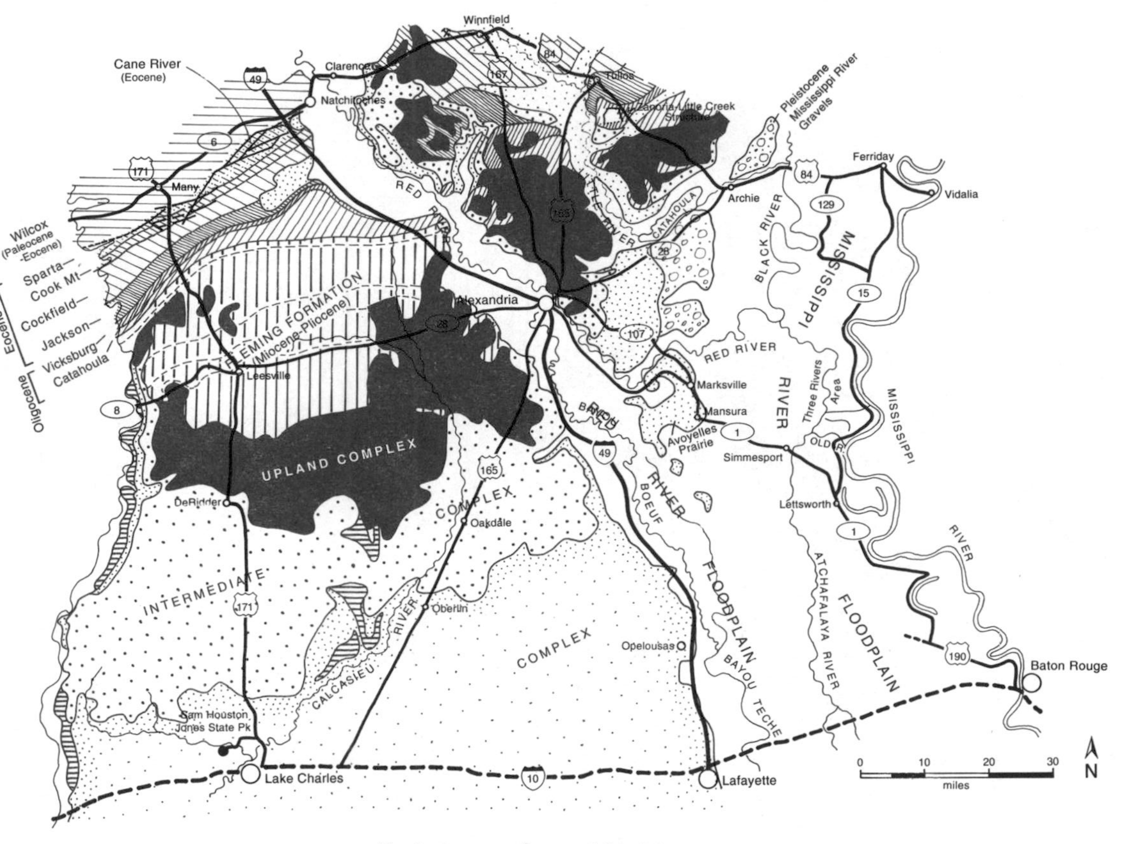

Geologic map of central Louisiana.

Central Louisiana

Great Rocks and Great Rivers

The Tertiary uplands of central Louisiana offer the widest variety of rocks in the state. Bands of sediment, mostly old river sands, delta muds, and swampy lignite, arc across the region. They stand out in outcrops and roadcuts, so you can actually see them from your car.

The sediment bands record the progressive growth of the coastal plain into the Gulf of Mexico as rivers dumped mud, sand, and silt where they reached the coast. The environments in which the sediments were deposited were similar to the environments we see along the Gulf Coast today: rivers, natural levees, deltas, marshes, swamps, floodplains, beaches, barrier islands, and shallow sea. A few beds of limy silt contain marine fossils that tell of times when shallow seawater flooded the lowlands.

The Red, Atchafalaya, Ouachita, and Mississippi Rivers converge in the Three Rivers area of east-central Louisiana. As early as 1830, river men and engineers cut navigation channels across the narrow necks of Mississippi River meander loops; people still manipulate the rivers. The principal control structures are easily visible from the roads in the Three Rivers area.

East of Alexandria, the Red River leaves its valley to shortcut to the Mississippi River through Moncla Gap, a narrow defile slashed through hills of Pleistocene deposits nearly 2,000 years ago. You can visit Moncla Gap on good roads a short distance from Alexandria.

Near Winnfield, you can drive over a salt dome and peer into a splendid quarry cut deep into its caprock. Near Ferriday are outstanding features of modern river and floodplain deposits. Sediments deposited during the ice ages make broad terraces that spread across the southern half of central Louisiana, overlapping older sediments to the north. They support profitable farms and provide pathways for roads and expressways. Many towns and cities stand on bluffs on the edges of these deposits, safely above flood level. Marksville, for example, is on a bluff

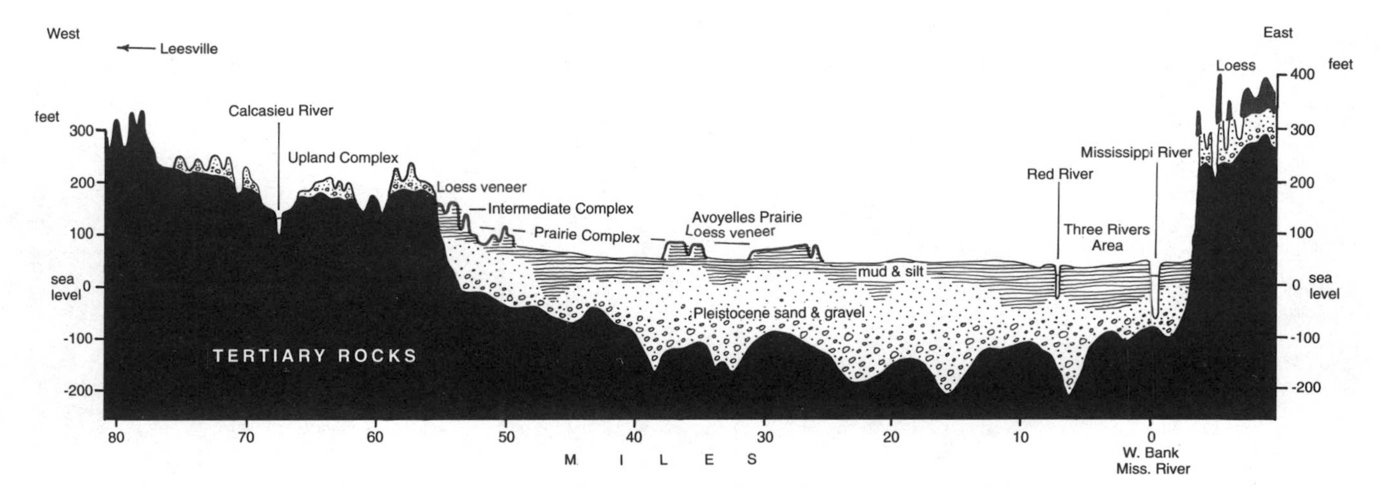

East-west profile from the Three Rivers area, through Marksville toward Leesville, showing dimensions and nature of fill in the Mississippi–Red River valley. —Saucier and Autin, 1990

at the edge of Avoyelles Prairie that Indians chose as the site of a town and ceremonial mounds.

Interstate 49
Alexandria—Opelousas—Lafayette
90 miles

Most of the highway follows Bayou Boeuf along the west margin of the Red River floodplain.

The Red River hugs the east bank of its floodplain southeast of Alexandria, then abruptly turns to the northeast, abandoning its own valley in favor of a shortcut to the Mississippi River through a narrow breach called Moncla Gap. The expanse of flat terrain stretching east to the horizon between Opelousas and the US 167 crossover to Meeker is the floodplain in which the Red River flowed before the diversion. The red farm fields are evidence of the former passage of the Red River. Geologists and archeologists think the Red River diverted through Moncla Gap about 1,600 to 2,000 years ago, judging from river terraces and pieces of Indian pottery they contain. Bayou Boeuf now follows the former channel of the Red River.

Bayou Boeuf is a puny remnant of the Red River. Indeed, it seems to have shriveled during historic time. Many plantations, such as Loyd's Hall, five miles south of Lecompte off US 71, were established along its banks because river steamboats plied the bayou, carrying sugarcane and cotton to market. Folktale claims the Lloyd's of London family set up a disinherited black sheep in America with the understanding that he change his name; he changed it to Loyd.

Between Washington, a few miles north of Opelousas, and Lafayette, the highway follows the edge of noticeably higher and hilly country of the Prairie Complex, which was deposited during Pleistocene time. Geologists have mapped abandoned river channels on this surface, remnants of a stream that flowed southwest. The channels contain red sand, which must mean that the Red River flowed through them. Watch for tan to red Prairie Complex sand quarries and roadcuts between Lafayette and Opelousas, especially at the southwestern corner of the US 167 crossover four miles north of Opelousas. Opelousas and Lafayette are built on the eastern edge of the Prairie Complex, securely above the floods yet close to the agriculturally rich floodplain.

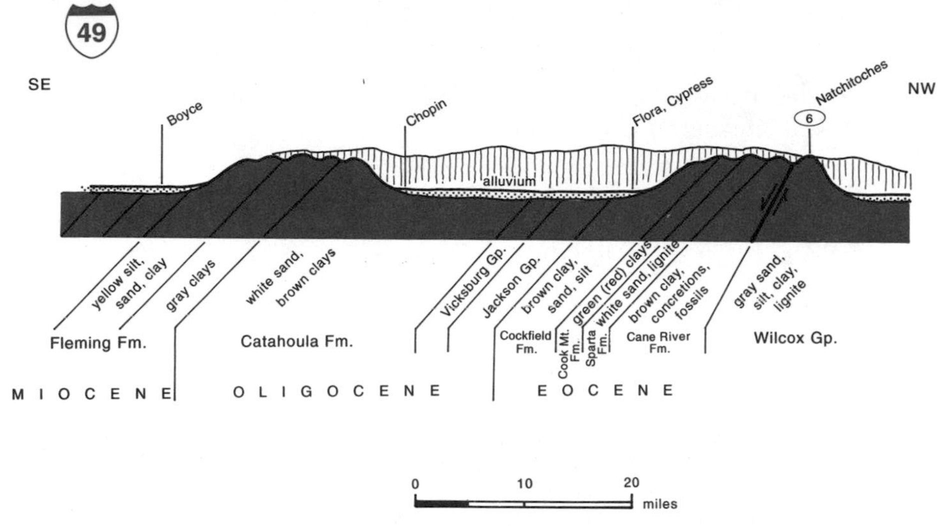

Cross section of the geology south of Natchitoches along I-49.

Interstate 49
Alexandria—Natchitoches
50 miles

This pleasant road crosses two hilly areas and two parts of the Red River valley. It crosses young floodplain deposits in the valley, and older rock formations in the hills. The older formations are exposed in broad bands that become progressively younger southward. The rocks are nicely exposed in the hills; you can see quite a variety of sandstones, shales in layers as thin as sheets of paper, and occasional concretions and fossil seashells.

The highway through downtown Alexandria passes close enough to the Red River to provide a view of the artificial levees on the west bank and Pleistocene bluffs on the east bank. The site of the Great Rapids of the Red River, for which Rapides Parish was named, is between the downtown bridges and the Highway 71 bridge. The rapids formed where the river cut across a resistant mudstone ridge in the Miocene Fleming formation. The rapids were the head of river navigation, which explains the locations of Alexandria and Pineville. Even though early river men cut shallow channels through them, low water usually made the rapids impassable to all but the smallest vessels.

During the Civil War, U.S. Navy gunboats were trapped above the rapids by a sudden drop in river level in the summer of 1864. After unloading armor and artillery, the big boats still had too much draft to float the falls, and the advancing Confederates threatened to capture them. A clever Yankee engineer dammed the river to flood the rapids with enough water to pass the gunboats, an old timberman's technique. Remnants of this old dam could be seen at low water until the Corps of Engineers channelized and dammed the river in the 1980s.

The highway crosses the floodplain of the Red River in the 13 miles between Alexandria and Boyce. It stays west of the river, which swings from the eastern side of its valley at Alexandria to the western side at Boyce. The floodplain is flat and its soils characteristically red, as everywhere along the Red River.

Two miles north of Boyce, the landscape abruptly rises into a tract of rolling hills covered with pine trees that continues for ten miles. The bedrock consists of formations deposited during Oligocene and Miocene time. The sandstones form hills because they resist erosion better than the mudstones and shales, which make up the lower areas. Pine trees like soils that provide both water and good drainage, the kind that generally develop on sandstone.

The Fleming formation, which was deposited during Miocene time, dips very gently down to the south in a band about eight miles wide. It overlies the much thinner Catahoula sandstone, which was deposited during Oligocene time, and is exposed in a band about two miles wide at the northern end of this set of hills. Look between Boyce and the Lena exit for exposures of white, orange, tan, and gray sandstone, all part of the Fleming formation, all stream deposits. The Fleming formation also contains mud and silt that were laid down in delta backswamps, but you rarely see them.

The white Catahoula sandstone makes a ridge just north of Lena, where the pine trees thrive. The Red River valley necks down to a narrow defile a few miles east of the highway, where the river cuts through the ridge of resistant Catahoula sandstone; roadcuts expose white Catahoula sandstone. Fairly large streams deposited the Catahoula sands, if we can judge from their thickness and wide lateral extent. Chunks of fossilized palm wood leave little doubt that the climate of Oligocene time was fairly tropical.

The interstate highway drops to river level at the Chopin exit, and crosses a broad expanse of Red River floodplain between Chopin and Cypress, 12 miles wide. The Catahoula sandstones in the hills to the west must be a little softer than those lower in the formation, because the river has cut them back to open this broad valley.

Another set of sandstone bands, these Eocene in age, are exposed in hilly ridges between Cypress and the Louisiana 6 crossover west of Natchitoches. The shale layers between beds of sandstone appear in several roadcuts, still not weathered beyond recognition.

The southernmost roadcuts in reddish sandstone in the hills north of Cypress expose the Cook Mountain and Cockfield formations. Brown, papery shale beds and thin layers of sand and silt at seven miles south of Louisiana 6 appear in very nice small exposures in the expressway median. These are at the top of the Cook Mountain formation, beneath the Sparta formation. The Sparta formation is at the surface, between two and four miles south of Louisiana 6; look for thin layers of dark lignite coal overlain by white sand. Iron oxides have stained the sand orange in places, making the white traces of fossil roots stand out. Watch about one and a half miles south of Louisiana 6 for low roadside exposures of the Cane River formation, thinly bedded gray shales in which the slope wash from heavy rains has cut an intricate pattern of rills. Marine fossils abound in these beds: snails, clams, corals, and oysters. They appear as bright white specks against the dark gray background of the shale.

A fault about a half mile south of the Louisiana 6 crossover separates the Eocene Cane River formation from the older beds of the Paleocene Wilcox formation. You can see the fault on geologic maps that cover a large area, but not from the highway. Nice roadcuts in the Wilcox formation expose shales, sandstones, and lignite coals. Some of the sandstones have eroded into hoodoos, those funny erosion pedestals that become animal and human caricatures if you stare at them for a minute or two.

Brown, papery claystone, capped by thin, concretion-laden sandstone (Eocene Cook Mountain formation), along I-49, 6.8 miles north of the Louisiana 120 crossover near Flora/Cypress.

Erosion rills in marine clays, along I-49 south of Natchitoches.

Fossil debris of snails, clams, corals, and oysters in gray marine clay beds (Eocene Cane River formation). West side of I-49, 1.5 miles south of Louisiana 6 crossover near Natchitoches. Quarter, top center, *for scale.*

US 84
Tullos—Jena—Archie
28 miles

The highway crosses at right angles the outcrop bands of the Eocene Jackson, Oligocene Vicksburg, and Oligocene Catahoula formations. The oldest are in the north, and they become progressively younger southward.

Upland Complex deposits cover most of the Catahoula rocks between Trout and Whitehall. The landscape along this stretch is pleasantly hilly; the dark green pines contrast with the orange soils, but few rocks are visible. Rocks in the Jackson and Vicksburg formations are mostly clay, which generally forms lowlands because it weathers and erodes so easily. The Catahoula formation is mostly sandstone and crops out more commonly than either the Jackson or Vicksburg formations.

Watch carefully for a yellow sand quarry east of the road five miles south of Tullos. The geologic map shows Vicksburg clay in this area, so what is yellow sand doing here? This is the corner of a large patch of sand that extends west of the highway. The sand is in the Fleming formation, which was deposited during Miocene time, and definitely does not belong within the outcrop area of the older Vicksburg formation. In their attempts to explain the situation, geologists have drawn a nearly circular fault system around the sand patch, and proposed that it dropped along the faults in some kind of collapse structure. The offset is about 2,000 to 4,000 feet, and the structure is still sinking. The cause of the collapse is a mystery.

Some geologists have suggested that a salt dome could have started here, then lost its supply of salt when another dome began to rise nearby, at the southern end of the North Louisiana salt basin. Other geologists suppose that a very deep mass of molten magma may have started the structure. Still others think a meteorite impact may have created a hole in which the sand accumulated. Whatever the explanation, the Fleming sand is good, and you can see it exposed in the pit.

White and tan sand of the Catahoula formation underlies orange soil in another quarry about a mile north of Trout, east of Highway 84. This exposure is in the lower part of the Catahoula formation, which disappears beneath Upland Complex deposits just south of the quarry. The red soil and dissected countryside of the Upland Complex continue south through Jena nearly to Whitehall. There, a narrow band of Intermediate Complex underlies the town and borders the modern floodplain between Whitehall and Louisiana 28.

84

US 84
Vidalia—Ferriday—Archie
36 miles

Highway 84 crosses the Mississippi River floodplain between Vidalia and Jonesville, at one of its narrowest spots. Bank to bank, the floodplain here is only 15 miles wide; in some other places, it is as much as 60 miles wide. The road follows high ground on natural levees all the way. Indians first used this narrow route on high ground—an eight-foot-high Indian mound at Frogmore tells of their passage. Nolan's Trace, a trail to Texas, used the same route in the late 1700s.

Between Vidalia and Ferriday, the highway follows the crest of a natural levee, which slopes gently away from the road for about a half mile on either side. It was above most of the floods, above the surrounding backswamps, which is why the older houses string along the road. In its early days, Vidalia linked the cattle trail from Texas with steamboats on the Mississippi River. The high levee north of the highway now protects the town and road from floods.

Watch along the four miles west of Ferriday for the rolling topography of ridges and swales on either side of the highway, especially to the north. These are sand ridges deposited on the point bar of an inside bend of the Mississippi River when it flowed across this area. Look for the Frogmore Indian mound and the Frogmore Cotton Gin, both at the junction of US 84 and Louisiana 566.

Highway 84 follows the crest of the natural levee between Frogmore and Jonesville, sharing the high ground with the railroads. Jonesville is on the Black River just below its origin in the confluence of the Ouachita, Tensas, and Little Rivers. The Black River flows into the Red River east of Moncla Gap. The road follows the Little River west of Jonesville to Archie, skirting the southern edge of a patch of braided stream deposits left as glacial outwash by the ancestral Mississippi River. It was probably part of Macon Ridge before the modern streams eroded their valleys through it.

84 6

US 84, Louisiana 6
Natchitoches—Winnfield Salt Dome—Winnfield
24 miles

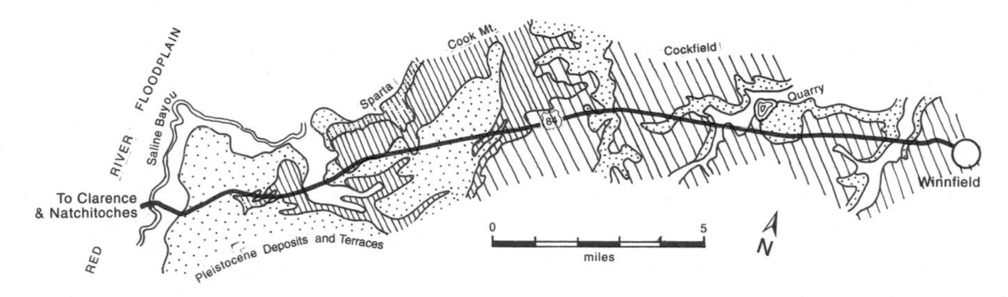

Geologic map along US 84 between Natchitoches and Winnfield.
—Belchic, 1960

This road amply proves that Louisiana really does have some genuine rocks. More rocky exposures and more rock types appear between Natchitoches (pronounced *NACK-o-tish*) and Winnfield than along any other road in the state.

Natchitoches is the oldest town in the Louisiana Purchase, settled by the French in 1714, and first named Fort St. Jean Baptiste. Later it was named for a Caddo Indian tribe. Early steamboats could navigate only as far as Natchitoches, where the great Red River log jam stopped them. The town began as a port town on the river, but lost its port status when the Red River changed course in 1834. The Cane River meander loop quickly silted up and thereafter flowed only during floods. The meander was dammed at both ends decades ago and is officially Cane River Lake, but everybody calls it Cane River.

Louisiana 6 heads north from Natchitoches on a bench of Prairie Complex that overlooks the Cane River. The road turns east about four

Grand Ecore, a beautiful bluff, exposes layers of the Wilcox formation at US 84 bridge over the Red River, north of Natchitoches.

Varicolored sandstone (Eocene Cockfield formation) roadcut 12 miles east of Clarence on US 84. Roots are preserved in the sand.

miles north of town to cross the Red River on a bridge that arcs out over the river from a high bluff on the western bank. The bluff is a famous historic landmark, known as Grand Ecore (Big Cliff). Impressive not only for its geologic exposure but also for its commanding view over the Red River, it has been fortified many times since the early Colonial period. It strategically overlooked the foot of the Great Raft, the Red River log jam, and the confluence of the Cane and Red Rivers.

If you are traveling east, you can easily miss the sheer face of Grand Ecore beneath the bridge. But a right turn at the abutment onto a short piece of old highway heads you back toward the river for a terrific view of the cliff. The river cut this impressive vertical wall out of a hill eroded into the Paleocene Wilcox formation. Layer upon layer of sand, silt, clay, and lignite coal are stacked in full view. This is the best natural exhibit of Wilcox rocks in Louisiana.

From the high bridge, you can see reddish sandbars in the river when the water is low. Red soil in the fields on the floodplain tells of floods on the Red River, spreading the red sediment of northern Texas and Oklahoma across Louisiana. Between Clarence and Winnfield, the highway follows the Harrisonburg Road, a historic route that began as an Indian trail. The road connected Natchez, Mississippi, and Nachitoches, Louisiana.

Watch for the levee 4.5 miles east of Clarence, and for the abrupt hills along the eastern edge of the floodplain five miles east of town. The road east to Winnfield crosses hills covered with pine trees.

Winn Rock quarry in caprock of Winnfield salt dome, west of Winnfield on US 84.

The Kisatchie National Forest boundary is 2 miles east of the river valley. A substantial roadcut in red sandstone of the Eocene Cook Mountain formation is north of the road. The small cuts in yellow to orange sandstone and soil at the tops of hills along the next 4 to 5 miles are exposures of Upland Complex deposits that cover the older Eocene rocks. The Cockfield formation, a deposit of Eocene time, appears in a large exposure of mottled yellow, pink, and red sandstone north of the road 12 miles east of Clarence. Look for remnants of white clay pebbles, root traces, and erosion rills in this river channel deposit. Hills of Pleistocene Upland Complex and Eocene Cockfield formation continue east to Winnfield.

Winnfield Salt Dome

About 20 miles east of Clarence, 4 miles west of Winnfield, watch north of the road for the deep Winn Rock quarry in the limestone caprock of the Winnfield salt dome. Blocks of gray-and-white-banded calcite stand out along the road to guide you to the quarry. The quarry is closed to visitors, but you can see it from a view box, where the wooden fence joins the chain-link fence north of the road, a few hundred yards west of the entrance road. You can look right into the quarry, which is about 100 feet deep. Big blocks scattered along the road provide a close view of the rocks.

Winnfield salt dome is one of 19 in the North Louisiana salt dome basin, one of the few with caprock at the surface. The Louann salt was deposited during Jurassic time, and began to rise up through a thick cover

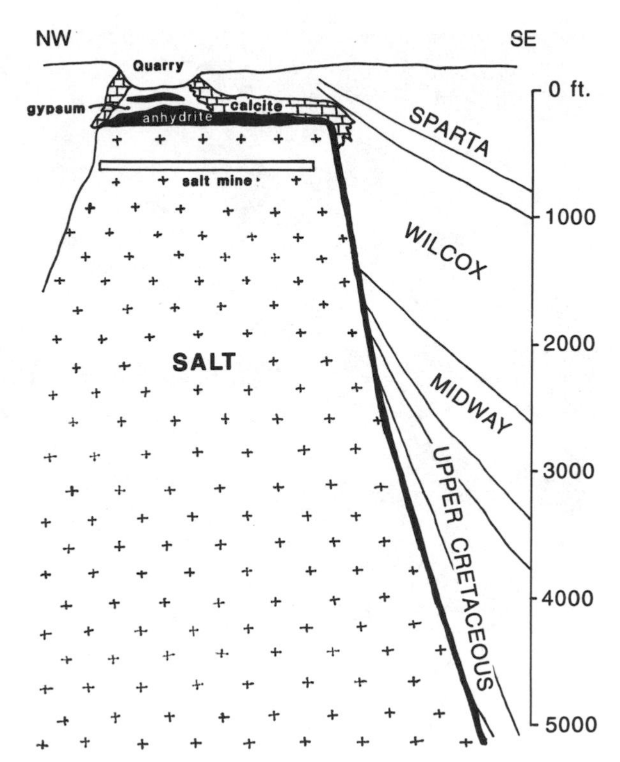

Profile of the Winnfield salt dome and quarry near Winnfield.
—Belchic, 1960

of heavier sedimentary rocks during late Jurassic time. It has now been rising for more than 100 million years. The salt at Winnfield dome rose to within 340 feet of the surface, raising a hill nearly 120 feet; you can drive up that hill to the quarry from either direction. Layers of Cockfield formation are tilted down to the west on the flank of the dome, about about one or two miles west of the quarry entrance. Limestone escarpments 60 feet high originally surrounded a pond in the central depression on the hill, and outcrops of steeply tilted layers of sandstone in the Cook Mountain, Sparta, Cane River, and Wilcox formations rimmed the dome. Spoil heaps now cover them.

The Winnfield quarry exposes three zones of caprock. The lower part is anhydrite—dark gray, massive, and crystalline. White-and-gray-banded calcite is exposed in the upper part of the quarry. Between them is a transition zone in which calcite and anhydrite are interbedded and mixed with gypsum, sulfide minerals, and minor amounts of other minerals. Most of the quarry blocks along the highway are banded calcite from the upper zone. The caprock lies on a flat surface of salt, the top of the giant pillar of salt that extends two miles down.

Block of calcite caprock along US 84 by Winn Rock quarry. Penny for scale at left of center.

Caprock is what remains after groundwater dissolves salt. It forms when the salt pillar has risen high enough to hit fresh water, which dissolves the salt, leaving the less soluble calcite, anhydrite, and gypsum behind.

Other minerals in the caprock include the iron sulfides pyrrhotite, marcasite, and pyrite, as well as the zinc sulfide sphalerite, and the lead sulfide galena. Another caprock mineral, barite, is barium sulfate, which because of its high density is used in making drilling mud. Geologists believe the metals probably came out of hot brines that rose from the depths along fractures in the rocks around the salt dome and reacted with sulphur in the caprock.

Winnfield dome was mined in the 1800s, mainly for building stone and to make lime. In 1930, the Carey Salt Company sank a mine 838 feet into the dome, 435 of those feet into salt, which was used in making chemicals, for curing meat, and for table salt. The extremely pure calcite in the caprock has been mined nearly continuously since 1936 for aggregate, and to make chemicals such as soda ash, caustic soda, and chlorine. The limestone is now exhausted and the mined anydrite fails to meet highway construction standards. The product is now mostly used for oil field roads, railroad ballast, and private driveways and parking lots.

US 84, Louisiana 34
Winnfield—Tullos
23 miles

The Eocene Cockfield formation is the bedrock between Winnfield and Tullos, but you see very little of it. The tan to yellowish tan sandy soils are characteristic. And the pine trees prefer sandy soils with good drainage, which the Cockfield formation provides.

The highway crosses the Dugdemona River and its swampy lowland east of Winnfield. Then it turns south at Joyce and crosses the Prairie Complex on the east bank of the Dugdemona River along most of the way to Tullos. Rolling hills between 7 and 11 miles south of Joyce indicate the road has climbed onto the Cockfield formation, above the lower and more subdued terrace level of the Prairie Complex. The highway crosses the Prairie Complex in the remaining 7 miles north of Tullos.

The road crosses Castor Creek, a tributary of the Dugdemona River, just north of Tullos. The town is on the northern edge of the outcrop band of the Eocene Jackson formation, which dips very gently to the southeast. The Jackson formation is composed of mostly fossiliferous gray and brown clay, with some ironstone layers and a few thin sand beds. Because of all the clay, the Jackson formation weathers easily and is rarely exposed.

US 165
Alexandria—Georgetown—Tullos
36 miles

The 20 miles of road north of Alexandria cross sediments of the Upland Complex, now eroded into low hills. Many small erosion cuts along this stretch expose sandy sediments in shades of tan and orange, the characteristic material of the Upland Complex.

The high bridge over the Red River in Alexandria gives a good view of the hilly country of the Upland Complex east of the river, the bends in the river, and the protective levees along its west bank. The bridge is directly over the old upper falls of the Great Rapids of the Red River, now drowned behind Lock and Dam #2. Hard, gray mudstones of the Miocene Fleming formation, exposed at low water in bluffs on the east bank, created the rapids. Bridge engineers used this hard ridge to support the pier pilings.

A dam on Rocky Bayou, a small tributary of the Red River, impounds Buhlow Lake and the recreation area east of the bridge. An unnamed oxbow lake along the road was created by the Corps of Engineers as they straightened the river for the Red River Waterway project.

Five miles north of Pollock, ten miles south of Georgetown, the road dips into the valley of Fish Creek, where you can see distinctive outcrops of white sandstone, the Oligocene Catahoula formation. Watch carefully for white rocks next to a utility line that cuts through the forest immediately south of Fish Creek. The Catahoula formation here is a white sandstone that lies above gray silty claystone. Crossbeds are clearly preserved in the sandstone.

Between the area five miles south of Georgetown and Tullos, the road rolls up and down across the valleys of Bear Creek, Dugdemona River, and Castor Creek, which slice through sediments of the Prairie Complex exposed on the hilltops. Georgetown stands on the higher elevations of the Prairie Complex. Tullos is at the northern margin of a band of rocks that trend from northeast to southwest, part of the Jackson group of late Eocene formations. All you see of them are patches of pale tan sand next to the road in Tullos. The formation typically includes clays interlayered with thin beds of lignite coal, sands stained red and brown by iron oxides, and fossil beds.

US 165
Alexandria—Oakdale—Iowa (I-10)
80 miles

Three broad bands of sediment—the Upland, Intermediate, and Prairie Complexes—sweep across this part of central Louisiana, more or less parallel to the coast. Each has its own slightly distinctive topography, soil colors, and plant cover. Highway 165 crosses all three.

Highway 165 arcs west of Alexandria, crossing Bayou Rapides about a mile southwest of its junction with Louisiana 1. The city and this segment of the road are on natural levee and alluvial deposits of the Red River valley. Bayou Rapides follows an abandoned channel the Red River once followed along the west side of its valley. Indian villages and then early plantations were built on the high ground of the natural levees along Bayou Rapides, a navigable waterway to transport cotton and sugar to the Red River, and then to markets in New Orleans. Flat floodplain, typical red soils, and tilled farmland comprise the landscape along about nine miles of road southwest of Alexandria.

The landscape abruptly changes where the road climbs out of the river valley onto tan sand and silt deposits of the Intermediate Complex, here a strip about two miles wide.

Between Woodworth and Longleaf, the road crosses the higher and older Upland Complex. Its rolling hills covered with pine trees and red soils are quite distinctive. This piece of the Upland Complex is part of a band about 20 miles wide that trends generally parallel to the coast from Alabama to Texas. The Kisatchie National Forest lies west of the road for the entire stretch between Woodworth and Longleaf, nearly coincident with the band of Upland Complex.

The road between Longleaf and Oberlin crosses a wide expanse of the younger Intermediate Complex, on both sides of Oakdale. The hills are more subdued than those in the Upland Complex, and the soils come in shades of orange and tan. The road follows the Calcasieu River southwest toward Calcasieu Lake. Like the Upland Complex, the Intermediate Complex is a broad band generally parallel to the coast. It can be traced all along the Gulf states. It is younger than the Upland Complex and records the next episode of coastal fill as rivers built seaward into the Gulf of Mexico during Pleistocene time.

Between Oberlin and I-10, Highway 165 crosses the Prairie Complex, the youngest of the three big bands of Pleistocene sediment. A mildly undulating landscape, light tan soils, and more deciduous than coniferous trees characterize the Prairie Complex. This late Pleistocene wedge of river, floodplain, and delta sediments is at its widest, 30 to 50 miles, in East Texas and central Louisiana. The modern Louisiana coastal plain laps onto the Prairie Complex about 20 miles south of Interstate 10.

US 167
Alexandria—Winnfield
40 miles

The Pineville expressway crosses the Red River, loops around the east side of Alexandria, and climbs across the Prairie, Intermediate, and Upland Complexes.

North of Alexandria, the highway winds through low and rolling hills mantled in orange and red soil, the typical terrain of the Upland Complex. The road crosses the band of Upland Complex for 23 miles, to a point 8 miles north of Dry Prong. You will see no roadcuts or outcrops, but flashes of sandy, orange soil in ditches and in small roadside scars betray the underlying Upland Complex sediments.

North of Dry Prong, the Upland Complex gives way to gray clays of the Oligocene Vicksburg formation in a band about a mile wide. Immediately north of these, it crosses four more miles of gray clays, these in one of the Eocene formations of the Jackson group. Both formations weather readily, and are very poorly exposed. Farther north, the road crosses a narrow stretch of Intermediate Complex south of Bear Creek, then descends to the Prairie Complex. It crosses more Prairie Complex north of Bear Creek.

Brown soils along the road between Bear Creek and Winnfield are developed on clays, silts, and sands of the older Cockfield formation, which was laid down in Eocene time. This progression in ages of the rocks illustrates how the coastal plain grew seaward, one formation after another.

US 171
Lake Charles—DeRidder—Leesville—Many
101 miles

The route between Lake Charles and Many is a perfect transect across the bands of sediment that have progressively filled the Gulf of Mexico since the dinosaurs vanished at the end of Cretaceous time. The layers of rock tilt very gently to the south, like a slightly tipped deck of cards. The oldest layers are at the north end of the route, around Many. With each mile southward toward Lake Charles, the ages of the stacked layers become younger.

The rolling hills between Many and Pickering, eight miles south of Leesville, are eroded in formations laid down between Eocene and Pliocene time, between 55 and 2 million years ago, in round numbers. The southern half of the drive, between Pickering and Lake Charles, crosses bands of sedimentary rocks deposited during Pliocene and Pleistocene time. Erosion has not yet carved these young formations into hills, so the landscape is much flatter than that on the older rocks farther north.

Between Lake Charles and the area two miles north of Gillis, Highway 171 crosses the flat and nearly undissected surface of the Prairie Complex. Tan silt, clay, and sand occasionally appear in very poor roadcuts. About two miles north of I-10, the road crosses English Bayou, where

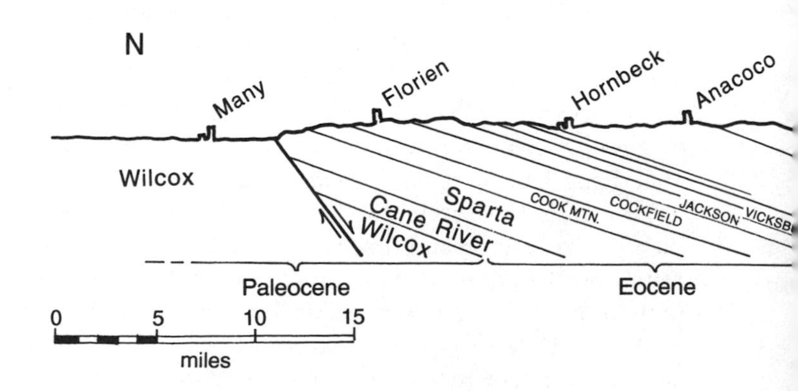

Geologic cross section along US 171 between Many and Pickering. The tilt to the rocks is greatly exaggerated in the diagram.

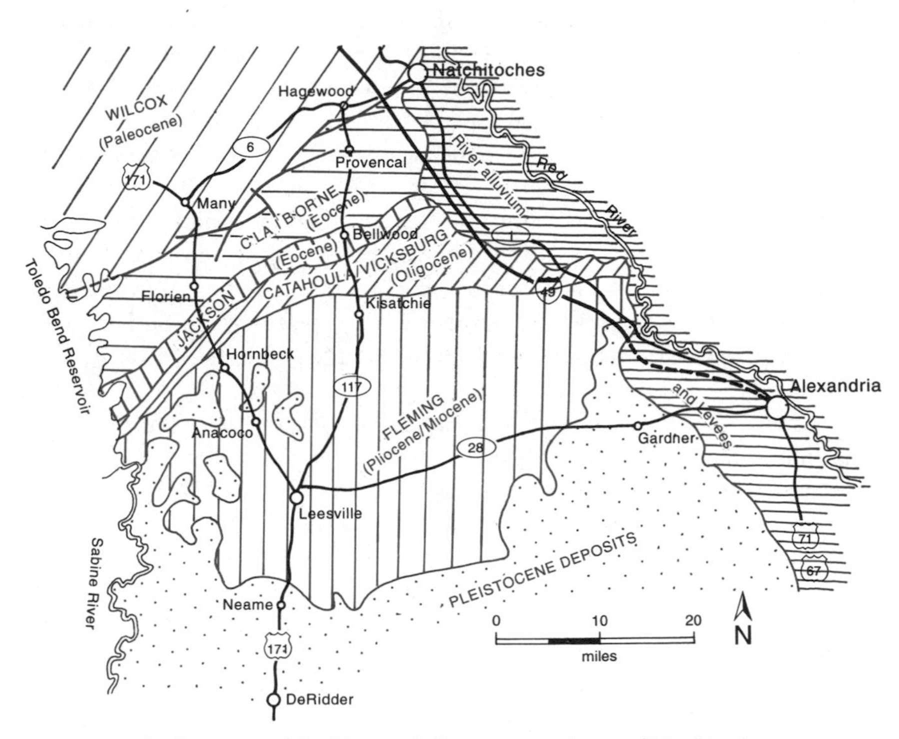

Geologic map of the Tertiary hill country southwest of Natchitoches.

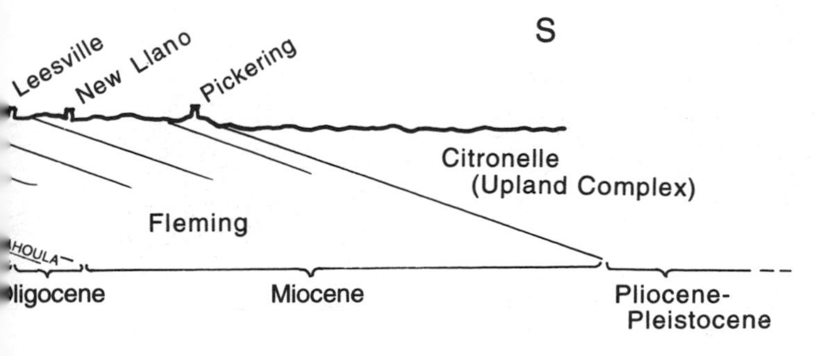

cypress swamps rim open water. The high bridge over the Calcasieu River is another mile north; you can look out over beautiful cypress swamps—once more extensive than now, to judge from the stumps. Just north of Moss Bluff, about four miles north of I-10, Louisiana 378 crosses US 171 and runs west for five miles to Sam Houston Jones State Park.

In the eight miles north of Moss Bluff, US 171 crosses the Prairie Complex. The road angles northwest from Gillis for two miles, and turns north where it crosses from Prairie to Intermediate Complex. The Intermediate Complex surface to the north is more rolling than the Prairie surface to the south. That is because the Intermediate Complex is older than the Prairie Complex, and has had more time to erode. Soils on the Intermediate Complex come in shades of orange and brown, a bit darker than those on the Prairie Complex.

The road through Ragley and Longville stays on the higher land of the Intermediate Complex between Barnes Creek to the east and Hickory Branch to the west. The highway joins Louisiana 26 east of DeRidder, and continues on the Intermediate Complex to Gillis.

Upland Complex deposits lie between DeRidder and Pickering, in a band about 12 miles wide. Note the steep slopes and dissected hills, as well as the darker orange to reddish brown soils so typical of the Upland Complex. Watch for bright orange and red bluffs in a cut on the west side of the road about 9 miles north of DeRidder. At 11.5 miles north of DeRidder, 2 miles south of Pickering, a red sandy stream channel of the Upland Complex east of the road is highly dissected into erosion rills. At road level, the channel cuts into stiff, gray floodplain clay of the Miocene Fleming formation.

Red, sandy stream channel rests on floodplain clays, two miles south of Pickering. Pliocene-Pleistocene Upland Complex deposit.

Local sandstone blocks (Fleming formation) used to build fence in Anacoco.

The southernmost outcrops of Tertiary rocks are around Pickering, where the upper, Pliocene, part of the Fleming formation is at the surface. You can see it in roadcuts at the junction of US 171 and Louisiana 10. Poor exposures of clay and sand persist to Leesville. About two miles north of Leesville, at the edges of Prairie Creek, light tan to white sandstone ledges of the middle part of the Miocene Fleming formation appear in roadcuts. The soil color is very pale to almost white. The white beds of limy rock along the roadside about five miles north of Leesville are in the lower middle part of the Fleming formation. At Anacoco, local fieldstone of hardened lower Fleming sandstone blocks and petrified wood are used in older houses and stone fences. Sandy exposures of lower Fleming formation exist north of Anacoco. A small mine near Anacoco produces opals.

The north edge of the outcrop band of Fleming formation extends two miles north of Hornbeck. A conspicuously white channel sand in the older Oligocene Catahoula formation is exposed west of the road about two and a half miles north of Hornbeck. The Catahoula formation is noted for its white sandstone ledges that crop out quite prominently in central Louisiana. The formation contains petrified palm wood.

Hodges Garden north of Hornbeck is a reclaimed quarry where hard sandstone from the Catahoula formation was mined for railroad ballast and used to build jetties at Galveston. The quarry has been transformed into a scenic garden with a lake, parks, hiking trails, and views. The rocks are nicely exposed in the old quarry walls, and in buildings and walls. The gardens are an example of how mined and exploited land can be restored.

White sandstone roadcut (Catahoula formation), 2.5 miles north of Hornbeck.

The highway crosses Toro Creek about five miles north of Hornbeck, where layers of the Oligocene Vicksburg formation lie south of the creek, layers of Eocene Jackson formation north of the creek. Both formations are mostly clay, easily eroded, and poorly exposed. It's hard to find good exposures of either.

The sandstone ridges that stand prominently around Florien are eroded in the Cockfield formation, which is also Eocene. Reddish hills three miles north of Florien are the Eocene Cook Mountain formation, which is just beneath the Cockfield formation. Look for a quarry in red Cook Mountain sandstone 3.6 miles north of Florien. Sparta sandstone is the oldest Eocene formation in the area. Look for its pale tan to white sandstone in roadcuts between 4 and 5 miles north of Florien.

Hilly terrain eroded in the Sparta sandstone abruptly meets flat country six miles north of Florien. The sharp boundary follows a fault that juxtaposes Sparta sandstone against the Paleocene Wilcox formation, which lies beneath the flat land north to Many. This fault, or system of faults, trends from northeast to southwest across central Louisiana, from Natchitoches to the Sabine River. It makes a sharp line in the landscape that separates hilly country to the south from flat country to the north. The fault surface dips down to the south, and the southern side moves down relative to the northern side. Geologists classify it as a normal fault.

Formations of the Wilcox group underlie the flat country around Many. Geologists divide the group into no less than ten formations, an assortment of sediments deposited in shallow seawater, and on deltas. The shallow marine sediments typically consist of sand, clay, and lignite coal deposited on a delta.

Louisiana 378
Sam Houston Jones State Park
5 miles

The park is along the banks of the Houston River, a tributary of the Calcasieu River. For part of its length, the park road parallels the Houston River, running between the river's natural levee and a watery backswamp that harbors a magnificent stand of cypress trees.

Its sandy hills give the park its distinctive character amid the surrounding flat land of the Prairie Complex. Look at the sand with a magnifying glass. The grains are almost all glassy quartz, almost all about the same size, and nicely rounded. It looks like the sand along the coastal beaches and barrier islands. The sandy hills in the park are at the eastern end of a long sand ridge that reaches to Galveston Bay near Houston, Texas.

Geologists call this sand ridge the Ingleside barrier trend. It lies on top of the Prairie Complex, so it must be Pleistocene in age. It was deposited when sea level was high during one of the warm interglacial episodes of Pleistocene time. This was the coastal barrier island about 135,000 years ago. The climate was warmer then, and sea level was higher. More water was in the oceans and less in the glacial ice of Greenland and Antarctica. If you can imagine gentle waves lapping against the sand ridges in the park, and envision Gulf of Mexico water extending south as far as your eye can see, you have the picture.

Sandy banks of the Houston River in Sam Houston Jones State Park. Low, sandy hills in park are remnants of a shoreline left over from the ice age.

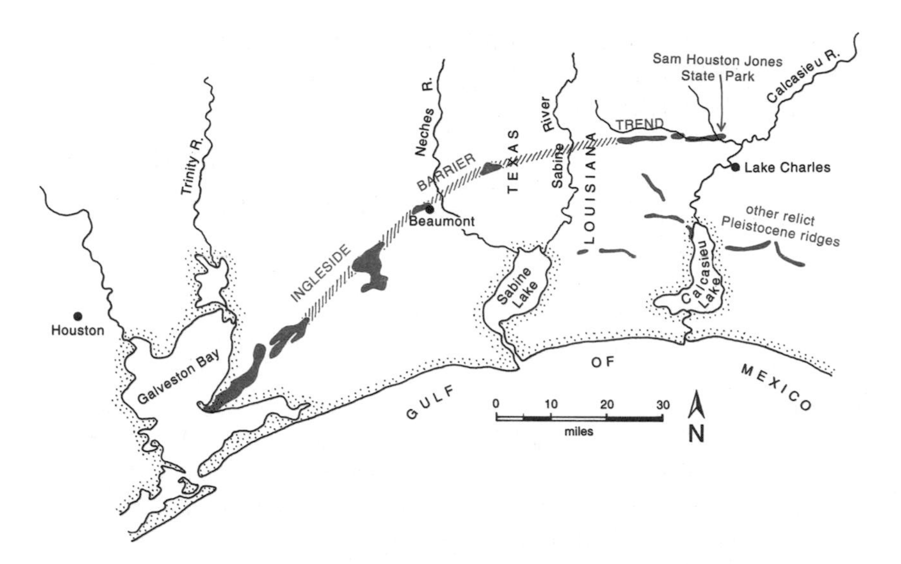

Map of the late Pleistocene Ingleside barrier trend. This relict sand shows where the Gulf of Mexico shoreline was about 120,000 to 130,000 years ago. —Saucier and Snead, 1989

False River (an oxbow lake) to right. Road runs on high natural levee and crevasse splay.

Louisiana 1
Baton Rouge—Simmesport—Marksville—Alexandria
115 miles

For an unhurried alternative to the freeway between Alexandria and Baton Rouge, try this bottomland run. The road samples the Mississippi, Atchafalaya, and Red Rivers, and the spillways, levees, and control structures along them.

East of Simmesport, the highway follows the Mississippi River in the inhabited upper part of the Atchafalaya basin. It crosses the Atchafalaya River at Simmesport, then cuts across the combined floodplains of the Mississippi and Atchafalaya Rivers to Avoyelles Prairie at Marksville. Between Marksville and Alexandria, the highway loops southwest off the prairie, and meets the Red River at the entrance to Moncla Gap, near Echo. Between Echo and Alexandria, the highway follows the natural levee on the southwest bank of the Red River.

In the Mississippi alluvial valley west of Baton Rouge, and in most of the valley, natural levee soils are light brown to gray in plowed fields. Swamp soils are dark gray to black.

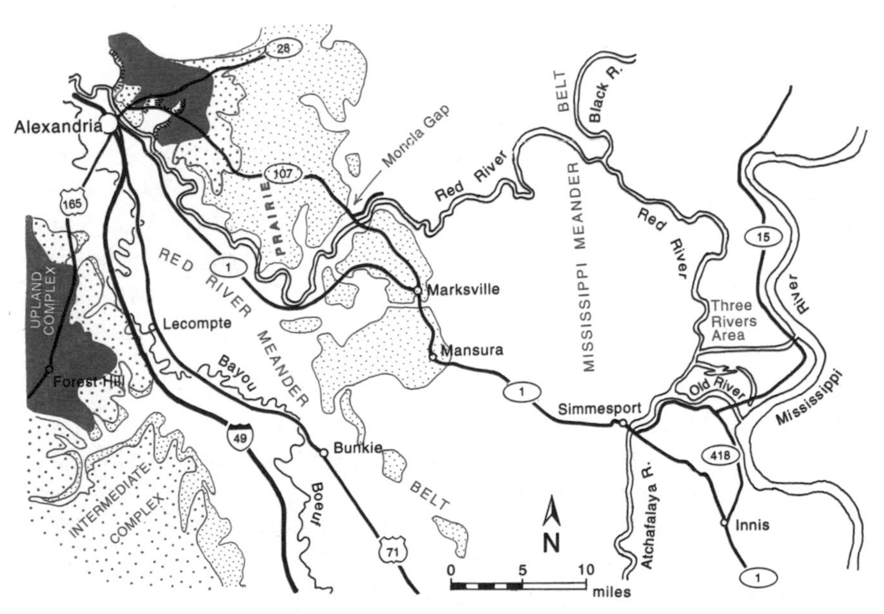

Geologic map of the area southeast of Alexandria.

The Parlange, an antebellum plantation house built on high ground along the False River.

You can reach Louisiana 1 from Baton Rouge by driving west on Airline Highway (US 190), or you can cross the Mississippi River on Interstate 10 and drive north three miles on Louisiana 415 at Port Allen to join US 190. The first few miles of US 190 track along a west loop of the Mississippi, where a grassed artificial levee dominates the skyline north of the highway. The road is on the natural levee; you can see the gentle slope north from the road toward the river loop.

About five miles north of US 190, immediately north of the junction between Louisiana 1 and Louisiana 416, the land rises abruptly as the road curves around the west side of False River Lake. The curved shape of the lake is typical of oxbow lakes that form as abandoned meander loops.

The road continues on land about 15 feet above the lake. Look west to see the land slope gently away from the lake. This broad patch of high ground is a crevasse splay. It formed when the Mississippi River breached its natural levee on the False River loop, spreading a thick carpet of mud and silt as the water velocity dropped immediately outside the breach. Antebellum homes and plantations built on this high ground still stand along Louisiana 1.

Ridge and swale topography on an old crevasse splay in the town of Labarre. Artificial levee rises in background.

New Roads was one the of earliest settlements in the Louisiana Purchase. West of New Roads, Louisiana 1 approaches another loop of the Mississippi, marked by the high levee north of the road. The levee is close to the road at Labarre, which is built on ridges and swales. Most houses stand on the ridges to avoid floods. The ridges and swales are part of a preserved crevasse splay that came off a meander loop of the Mississippi River.

At Morganza, the road closely skirts the levee to the north. Just west of town, Louisiana 1 climbs onto the levee and in a few miles crosses the Morganza spillway, an impressive Mississippi River control structure built

Louisiana 1 crosses over the Morganza spillway, built in 1954 and only opened once (1973) to let excess Mississippi River floodwaters spill into the Atchafalaya basin.

Conical burial mound (A.D. 1 to 400) characteristic of early Native American Marksville culture. Marksville State Commemorative Area.

by the Army Corps of Engineers in 1954. The elevation provides good views of the levee system and the river with its barges and ships. Watch from the spillway for the broad area of flat ground to the north, between the river and the spillway, where great volumes of water can be ponded in times of flood. When the water cannot be contained, the Morganza spillway releases Mississippi floodwater southwest into the Atchafalaya basin, where it can safely spread. The gates were opened during the great flood of 1973, but not during the flood of 1993.

Louisiana 1 follows the natural levee along the west bank of the Mississippi River between Morganza and Simmesport. At Lettsworth, Louisiana 15 heads north from Louisiana 1, to the Three Rivers area and the Old River control structures.

From the high bridge over the Atchafalaya River at Simmesport, you can see control levees along both sides of the river. Levees actually surround the town. The junction of the Red, Atchafalaya, and Old Rivers is only a few miles north of Simmesport. You can see the influence of the Red River in the red soils in fields west of the Atchafalaya River.

Eight miles west of Simmesport, the highway crosses the large, western control levee of the Atchafalaya basin. At Moreauville, on the western edge of the Atchafalaya basin, Bayou Moreau is an old Red River course, told by the reddish brown color of natural levee ridges.

The landscape changes near Mansura, as Louisiana 1 crosses from the lowlands of the Atchafalaya basin onto an isolated remnant of the Prairie Complex, the Avoyelles prairie. Like Indian tribes before them, settlers found the high edge of this prairie, overlooking the adjacent fertile floodplain, an excellent place to live and farm.

Houses in the French Colonial style, dating to the late 1700s, survive in Mansura, which is also renowned for its *cochon de lait*, roast suckling pig. The ups and downs in the road near Mansura are evidence of Pleistocene Mississippi River meander loops and point bars preserved on top of the Prairie Complex. They speak of the higher level of river flow in times past. Nearly three feet of loess covers the Avoyelles prairie surface in some places.

The road loops its way through Mansura, Marksville, and Echo, staying on the high ground of the Avoyelles prairie. Marksville is near an ancient walled and mounded city, built by Native American mound-builder people around 2,000 years ago. It is now preserved in the Marksville State Commemorative Area. The Tunica-Biloxi Indian Reservation, with its very nice museum and visitor's center, is also at Marksville.

At the town of Fifth Ward, Louisiana 1 leaves the Avoyelles Prairie, descends to river level in Moncla Gap, where the Red River jogs northeast to make its shortcut run to the Mississippi River.

The road between Echo and Alexandria trends northwest along the Red River on the west bank natural levee. Highlands to the northeast, across the river, are remnants of Pliocene and Pleistocene deposits of the Upland, Intermediate, and Prairie Complexes.

Louisiana 6
Sabine River—Many—Natchitoches
42 miles

Louisiana 6 follows the old Camino Real, a Spanish colonial road that linked Mexico City with Texas and Los Adaes, the easternmost Spanish post and mission a few miles west of the French fort at Natchitoches. Colonial powers contested the land between the Sabine and Red Rivers for more than a century.

Formations of the Paleocene Wilcox group are the bedrock along the entire route. But the Wilcox group is a complex sequence of delta deposits that vary greatly in color and include a wide assortment of sediment types. They reflect in their variety the range of depositional environments that exists on a delta.

Geologists have attempted to divide the Wilcox group into meaningful packages of rock, about ten different formations. But it is hard to trace them around central Louisiana. They are discontinuous, as you would expect on a delta where sedimentation is not uniform for any great distance. Marine clays abound in the Wilcox group, as do nonmarine beds of sand and clay that were laid down in distributary streams and floodplains. Beds of lignite coal come and go, showing that vegetation was as abundant during Paleocene time as now. A thick coal seam in the Wilcox group is mined in the Dolet Hills, near the junction of I-49 and US 84.

Fort Jesup was a military fort from 1820 to 1847. At the Fort Jesup State Commemorative Area east of Many, look in the foundations of the buildings for the blocks of pebbly sandstone with their stain of brown iron oxide. The blocks, sandstone probably from the upper part of the Wilcox group of formations, were gathered locally.

Good exposures of rocks in the Wilcox group are on either side of I-49, where Louisiana 6 crosses the freeway west of Natchitoches. Look for beds of sand, silt, and clay in various colors, and an occasional dark bed of lignite coal.

Louisiana 8/28
Alexandria—Leesville—Sabine River
48 miles

Most of the route crosses the Miocene Fleming formation, covered by Pleistocene Upland Complex deposits in the eastern part. The countryside is pleasantly rolling, particularly near stream cuts. A few good roadcuts provide glimpses of the rocks.

In the nine miles west of Alexandria the road crosses floodplain and natural levee deposits of the Red River valley. Watch for the characteristic red soil. Bayou Rapides parallels the highway about a mile to the north; its natural levee extends almost to the road. Indians and white settlers lived and farmed on its high ground.

The valley margin is easy to spot as the road climbs an abrupt hill, ascending through very poorly exposed gray silt and clay of the Williamson Creek member of the Fleming formation. It tops out on red sandy hills of the Upland Complex deposits that blanket the Miocene Fleming formation.

A gap in the hills offers a quick glimpse of Kincaid Lake to the south. A few miles north of the road is Hot Wells, known for its medicinal hot springs. The road crosses about ten miles of Upland Complex through Gardner to Cypress Bayou. Nearly five miles of terrace with buff and orange soil lie between Cypress Creek and the Calcasieu River. Gray soils west of the Calcasieu River as far as Leesville betray the gray silts and clays of the Castor Creek member of the Fleming formation. The gray clays weather easily, so you never see them well exposed in outcrop, but the gray soils are distinctive, quite different from the red soils of the overlying Upland Complex.

The area east of Leesville is the headwaters of the Calcasieu River. It becomes a major river as it picks up tributaries, then flows past Lake Charles on its way to the Gulf of Mexico.

Leesville is on hills eroded in the Fleming formation near Bayou Castor. A few roadcuts in town expose reddish sand terraces deposited during Pleistocene time. In the three miles west of Leesville, the road crosses gray soils developed on the Fleming formation. Farther west, Louisiana 8 crosses Prairie Creek, then follows red ridges in the Upland Complex toward Burr Ferry and the Sabine River. Watch for a nice roadcut on a ridge about 6.5 miles from the Sabine River, 11.5 miles west of Leesville. You can see yellowish root traces in the fine red sand of the Upland Complex, as well as spots of white clay that are probably weathered pebbles. A high vantage point three miles east of the Sabine River provides a very nice view through the trees to where Prairie Creek was dammed to make Anacoco Lake.

The road drops to the Intermediate Complex east of Burr Ferry, and to the Prairie Complex west of Burr Ferry. The Prairie Complex extends nearly to the eastern end of the Sabine River bridge.

Louisiana 15

Ferriday—Three Rivers Area—Old River Control Structures—Lettsworth

60 miles

Here in the Three Rivers area, remote wilderness is juxtaposed with the Old River control structures, certainly among the largest engineering projects in the world. No other part of Louisiana better epitomizes the interaction between people and rivers. Louisiana 15 provides the access through a scenic fabric woven of natural and cultivated backswamps, oxbow lakes, natural levees, and monumental artificial levees. Parts of the road are on the levees to stay out of reach of swampy land and floodwater.

At one time, the Red River and the Atchafalaya River did not connect with the Mississippi River where they do today. About 2,000 years ago, a small tributary that flowed out of Moncla Gap in the Avoyelles Prairie was overwhelmed by the Red River. The Red River abandoned its valley and took this new, shorter course to join the Mississippi River at the top of Turnbull's Bend. The Atchafalaya River was a small side stream that drained the western edge of the Mississippi Valley. As the Atchafalaya River eroded headward and Turnbull's Bend migrated laterally, the two eventually joined;

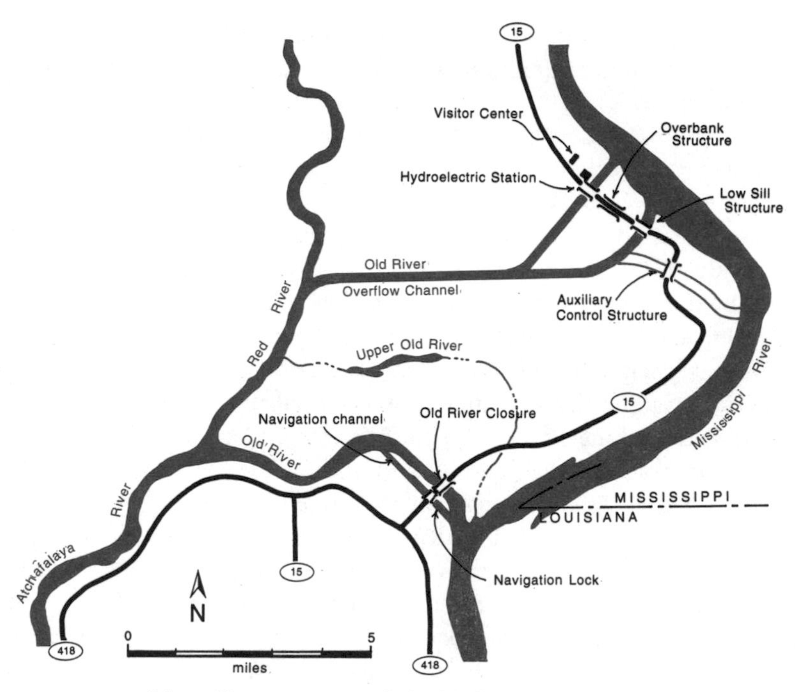

Map of Louisiana 15 and the Old River control structures.

the Atchafalaya River intersected the bottom of Turnbull's Bend to become a small distributary of the Mississippi River.

As steamboat captains well knew, long meander loops such as Turnbull's Bend added many miles to the length of the river. In 1831, the most famous of these boatmen, Captain Henry Shreve, dug a shortcut ditch across the narrow neck of Turnbull's Bend. As the northern loop of the bend dried up, the southern loop expanded to become the Old River, where westward flow from the Mississippi predominated, though occasionally floods on the Red River reversed the flow.

The Red and Mississippi Rivers now shared a new distributary channel, the Atchafalaya River, which was choked with logs. Pioneering attempts in the early 1800s to remove the 30 miles of jammed logs failed. In 1839, the state of Louisiana appropriated funds to clear the jam.

As water from the Mississippi River flowed west through Old River into the newly cleared Atchafalaya River, it widened and deepened. Meanwhile, water from the Red River flowed down the enlarged channel of the Atchafalaya River, never again able to overpower the westward flow in the Old River. Louisiana river watchers recognized that the Atchafalaya River

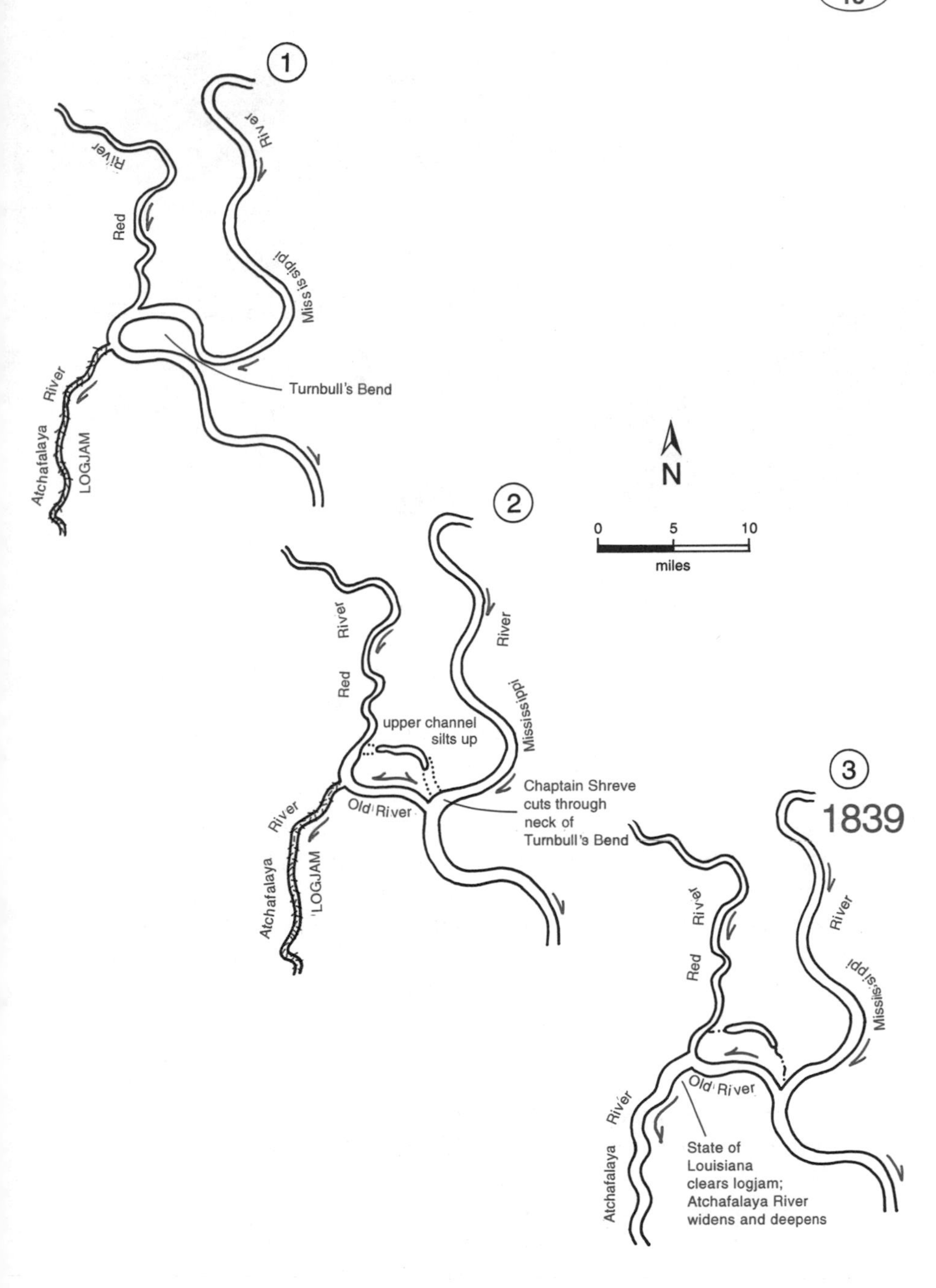

History of the Three Rivers area, showing how the Mississippi River captured the Red River at Turnbull's Bend, and how human intervention further altered the rivers.
—J. I. Snead, 1994, personal communication

Old River lock—navigation link between the Mississippi and Red Rivers.

could one day become the new path of the Mississippi River to the Gulf of Mexico.

Congress assigned the U.S. Army Corps of Engineers the task of improving navigation on the Mississippi River in 1879. By 1951, it was clear that the Mississippi River was imminently prepared to trade its current 315-mile course to the Gulf of Mexico for the 142-mile route through the Atchafalaya River.

To allow the Mississippi River to abandon its present channel and divert down the Atchafalaya River would lead to the abandonment of the vast infrastructure of towns, industry, and river traffic below Baton Rouge. It would also decrease the supply of fresh water as seawater flooded the former main channel. The inevitable degradation of the wild environment of the Atchafalaya basin was not a prime consideration during the 1950s.

The Mississippi River Commission, after studying the Atchafalaya problem, authorized construction of flood control structures at Three Rivers. An earthen dam was built across the Old River in the 1960s, along with a ship channel and navigation lock. Upriver, an artificial control channel and structure was dug to connect the Mississippi River with the Red and Atchafalaya Rivers. It was designed to divert 30 percent of the Mississippi River flow down the Atchafalaya River.

The Mississippi River supplies two-thirds of the water in the Atchafalaya River, the Red River the other third. But one-half of the sediment in the Atchafalaya River comes from the Red River. It drains the dry country of northern Texas and Oklahoma, where the scanty plant cover permits rapid soil erosion. Ongoing lock and dam construction on the Red

Auxiliary control structure—built in the 1980s after the great flood of 1973.

River, until recently America's largest undammed river, will reduce the amount of sediment delivered to the Atchafalaya River.

The Old River lock, channels, and flood control structures performed as designed until the flood of 1973 threatened to destroy the entire diversion structure. The Mississippi River nearly claimed its shorter path to the sea through the Atchafalaya River in one flood. The low-sill structure was badly undermined by swirling, turbulent waters that shook its steel walls. Shortly thereafter, the low-sill structure was reinforced, and an auxiliary control structure and flood channel were added in the 1980s to shore up the defenses.

Sydney A. Murray, Jr. Hydroelectric Power Station and debris dredged from outflow channel.

In the 1970s, the city of Vidalia adopted a plan to generate electric power; construction of Louisiana's first hydroelectric station on the Mississippi River began in 1985 as part of the Old River control complex. A prefabricated power plant equipped with eight horizontal turbines designed to work with a water drop of less than 25 feet was barged up the Mississippi River and installed in a new channel that connects the Mississippi River to the Old River outflow channel. Each turbine generates 24 megawatts. Electricity began to flow in 1990. Louisiana 15 passes over the lock, dam, control structures, and power station, giving you a spectacular view of the entire complex. The visitor's center just north of the power station displays models of the Old River control complex.

Louisiana 28
Alexandria—Walters—US 84 Junction
32 miles

This route steps down eastward across terraces between the floodplains of the Red and Mississippi Rivers as though they were stair steps. The Pineville Expressway loop east of Alexandria is on the highest step, the Upland Complex. The pine forest, red soils, and fairly steep stream valleys are quite characteristic. The Upland Complex deposits extend east to Libuse. East of Libuse, the road crosses three miles of lower and flatter Intermediate Complex. Then it descends to the even lower and flatter level of the Prairie Complex, which extends east six miles to Saline Bayou.

A distinct break in elevation defines the east side of the Prairie Complex at Saline Bayou. The road between Saline Bayou and Walters crosses sand and gravel deposits that form the southern tip of Macon Ridge. West of Walters, these gravelly sediments are at backswamp levels and do not make ridges as they do farther north. Catahoula Lake, on the swampy surface of the modern alluvial plain, is north of the road. It is a rich biological area now preserved as the Saline Wildlife Area.

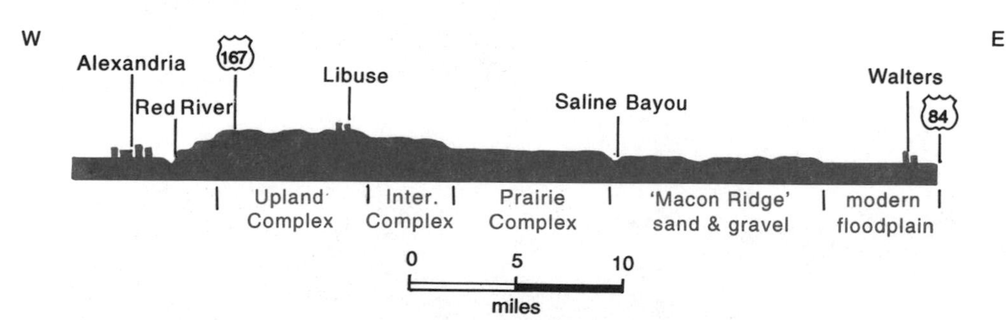

Profile of Pleistocene deposits and terrace levels between Alexandria and Walters.

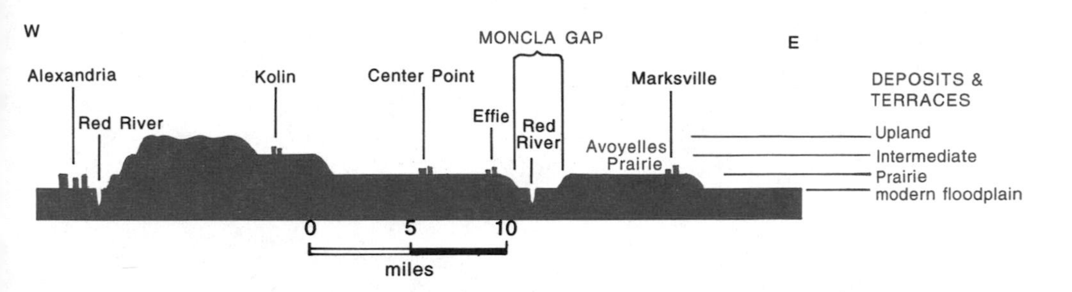

Profile of Pleistocene deposits and terrace levels along Louisiana 107 between Alexandria, Moncla Gap, and Marksville. The isolated patch of Prairie Complex called the Avoyelles Prairie, on which Marksville stands, is noticeable in this profile.

Louisiana 107
Alexandria—Moncla Gap—Marksville
30 miles

This short drive through the rolling hills east of Alexandria provides a view of the Red River's natural shortcut through Moncla Gap to the Mississippi River.

East of Alexandria, the highway quickly gains elevation in Pineville, across the river, as it ascends narrow bands of the successively higher Prairie, Intermediate, and Upland Complexes, which comprise the steep east bank of the Red River. Louisiana 107 crosses about five miles of piney red hills eroded in the Upland Complex, about 150 to 160 feet above sea level. This area is the highest for many miles around Alexandria.

The highway drops abruptly to the Intermediate Complex a mile west of Kolin, and crosses it for four miles. It drops to the flat Prairie Complex a mile west of the junction of Louisiana 107 and Louisiana 454, and crosses it in the eight miles through Center Point and Effie. You see no roadcuts and outcrops of the Prairie deposits, but can see the elevation changes from one terrace level to another.

Just east of Effie, Louisiana 107 leaves the Prairie Complex and descends to the floodplain of the Red River, in Moncla Gap. You can follow Louisiana 1196 to the U.S. Army Corps of Engineers Lock and Dam, and the Ben Routh Recreation Area. The park is a good vantage point to look out over the Red River and to see how narrow the valley is, about two miles wide. Elsewhere, the Red River valley is generally about ten miles wide. Moncla Gap is indeed narrow, which suggests that the Red River has not been flowing through it long enough to cut its normal ten-mile swath.

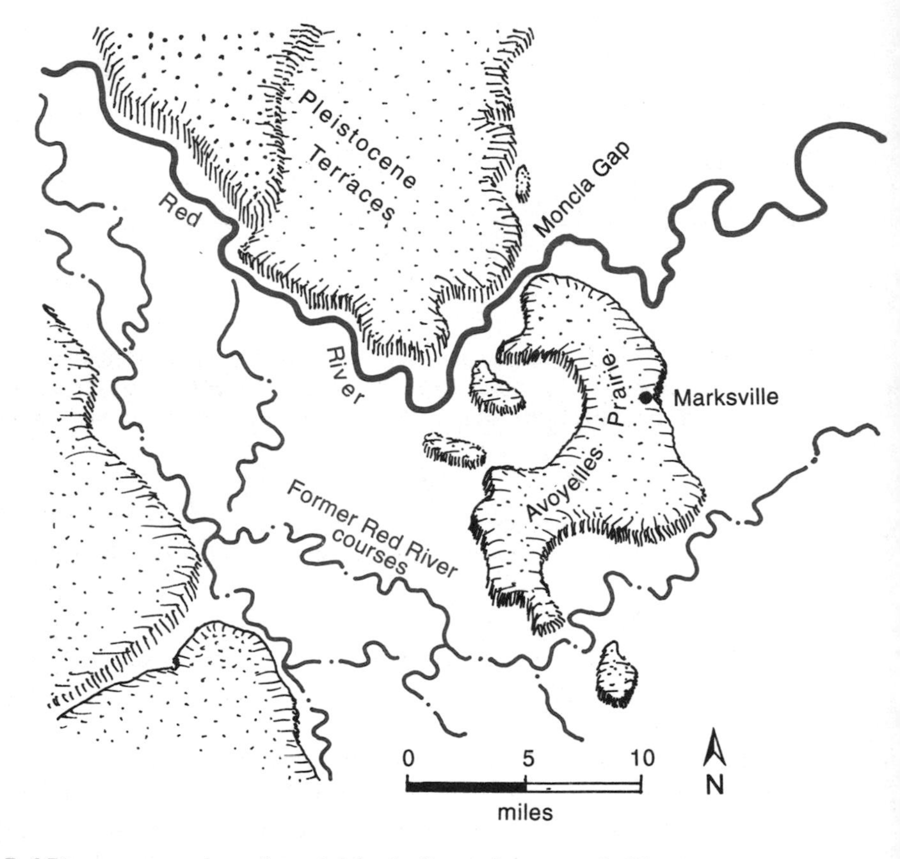

Red River turns northeast through Moncla Gap, a shortcut to the Mississippi River. The former Red River courses flowed around Marksville (on the Avoyelles Prairie) to get to the Mississippi. —Pearson and Hunter, 1993

Moncla Gap is also odd in trending northeast, almost at right angles to the normal southeastern trend of the Red River. What happened?

Geologists and soil scientists have studied the river and terrace deposits in the Moncla Gap, while archeologists have studied datable artifacts associated with those deposits. Recent studies of pottery fragments in the earliest natural levee deposits show that they were made by early Marksville people, the people who built the village and mounds in the Marksville State Commemorative Area. That dates the diversion of the Red River through Moncla Gap at a little less than 2,000 years ago.

Though it appears a bit odd at first glance, the diversion through Moncla Gap was a significant shortcut to the Mississippi River. It reduced by 32 miles the distance the Red River flowed to reach the level of the

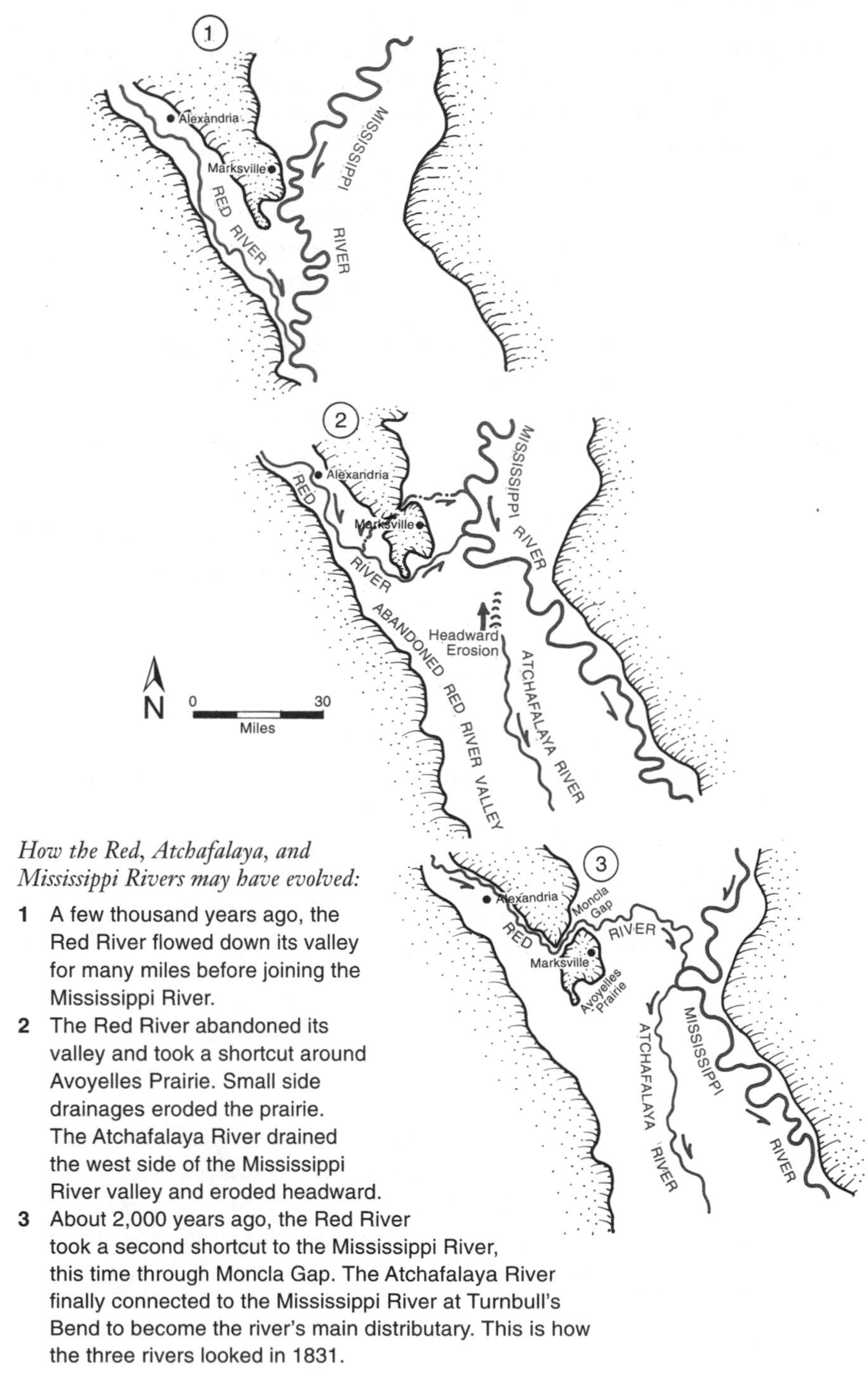

How the Red, Atchafalaya, and Mississippi Rivers may have evolved:

1. A few thousand years ago, the Red River flowed down its valley for many miles before joining the Mississippi River.
2. The Red River abandoned its valley and took a shortcut around Avoyelles Prairie. Small side drainages eroded the prairie. The Atchafalaya River drained the west side of the Mississippi River valley and eroded headward.
3. About 2,000 years ago, the Red River took a second shortcut to the Mississippi River, this time through Moncla Gap. The Atchafalaya River finally connected to the Mississippi River at Turnbull's Bend to become the river's main distributary. This is how the three rivers looked in 1831.

—Saucier and Sneed, 1989

Mississippi River floodplain. But why did the Red River cut through the barrier of Prairie Complex deposits here instead of finding a path around the Avoyelles Prairie? The map of old Red River courses shows that it did at one time flow around the end of Avoyelles Prairie. About 2,000 years ago, small streams had probably eroded headward across the prairie from both sides to create a gap. Then, probably during a very high flood, Red River water poured through, finding a shorter path to the Mississippi River. Some geologists think a fault or joint system trending northeast may have helped.

East of Moncla Gap, Louisiana 107/115 climbs to Marksville, at the high point of Avoyelles Prairie, the piece of Prairie Complex isolated by the Red River and Moncla Gap. Indian mound builders located here on a bluff overlooking the Old River about 2,000 years ago.

Louisiana 129

Features of the Floodplain

12 miles

If you see the Mississippi River and its floodplain from the air, the oxbow lakes, abandoned channels, crevasse splays, and backswamps are wonderfully laid out in plain sight. These features are difficult to see and interpret from the road because their lateral dimensions are measured in miles and their vertical dimensions in a few feet. On this short drive along Louisiana 129 it is possible, with a discerning eye and the help of a map, to see a good display of the Mississippi River floodplain. Numbered spots on the accompanying map help you locate the best places to see the most easily visible features.

West of Ferriday, US 84 follows the crest of an abandoned natural levee that once bordered the Mississippi River. The land is about 30 to 60 feet above sea level, and slopes a few inches per mile toward the Gulf of Mexico. The clay, silt, and fine sand of the floodplain were laid down within the past 6,000 years.

Four miles west of Ferriday (location 1), the road curves around the southern end of an abandoned channel of the Mississippi River, now a swamp about 3,000 feet wide. It is north of the road. Right along the northern edge of the road is a scarp 10 to 15 feet high that was once the steep bank of the outside bend of the channel. Highway 84 follows the swampy channel for several miles toward Stacy.

At Stacy, Louisiana 129 turns south from US 84. Three miles from US 84, Boggy Bayou (location 2) occupies an abandoned channel nearly filled with sediment. The old channel is the bayou, filled with swampy vegeta-

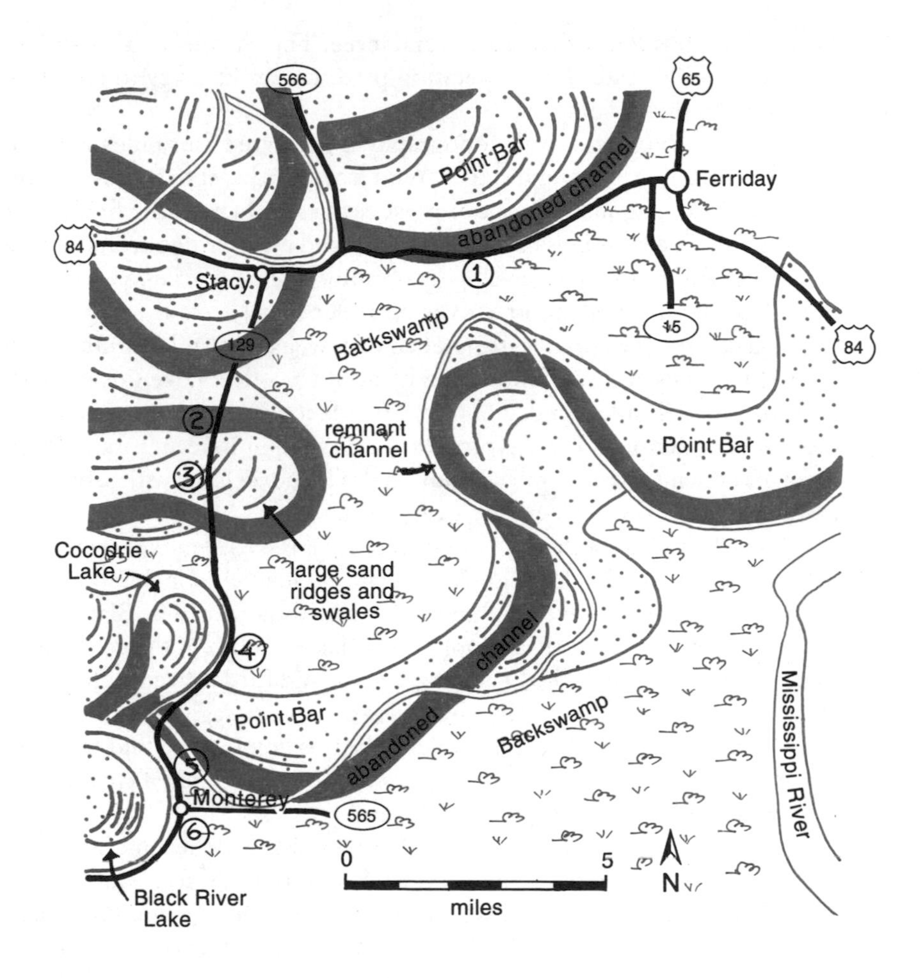

Geologic map of river environments on the floodplain southwest of Ferriday. Circled numbers refer to locations explained in the text. —Saucier, 1986

tion and five feet lower than the natural levee. The channel is fairly subtle, but the stop illustrates how vegetation patterns can be a key to relic features on the floodplain.

Excellent ridge and swale topography is preserved four miles south of US 84 (location 3). The sand ridges trend from northeast to southwest and cross the road diagonally. You can most easily see them in the fields west of the road. The buildings east of the road stand on the ridge crests. These dry ridges and wet swales developed as sandy sediment accumulated on a point bar, the inside bend of a river meander loop.

You can see the curving shape of an oxbow lake 7.5 miles south of US 84, where Louisiana 129 follows the bend of Cocodrie Lake (location 4). This oxbow is 4 miles long, 2,000 to 3,000 feet wide, and about 30 to 50 feet deep, dimensions fairly typical for a Mississippi River meander loop. A small remnant channel still actively feeds Cocodrie Lake with sediment, making it shallower and smaller every year.

Just 10.5 miles south of US 84 (location 5), the road curves around an oxbow lake of the Black River. A navigation channel amputated a large meander of the Black River, making it into an oxbow lake. The Black River occupies an abandoned Mississippi River channel south of the confluence of its three tributaries—the Ouachita, Tensas, and Little Rivers. The Black River flows into the Red River, which also occupies a segment of the same abandoned Mississippi River meander belt. It is common to find tributaries of the Mississippi River flowing in one of its abandoned channels as they make their way across its floodplain.

Watch for the small channel 0.2 mile south of the junction with Louisiana 565 at Monterey (location 6). It is about 100 feet wide and 10 to 15 feet deep. This is a crevasse channel, a place where a flood breached the natural levee beside the Black River oxbow when it was still a meander loop of the river. Water ran east through the broken levee, eroded this narrow channel, then dumped its load of sediment as it spread across the adjacent floodplain. The water deposited a crevasse splay deposit in a broad fan that covers about half a square mile. It is hard to see from ground level.

Location 3. Sandy ridges and watery swale between ridges, deposited originally on river point bar.

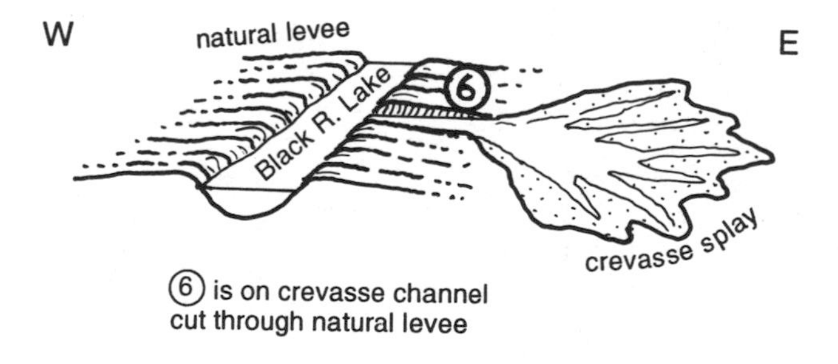

Location 6 is on a crevasse channel cut through a natural levee.

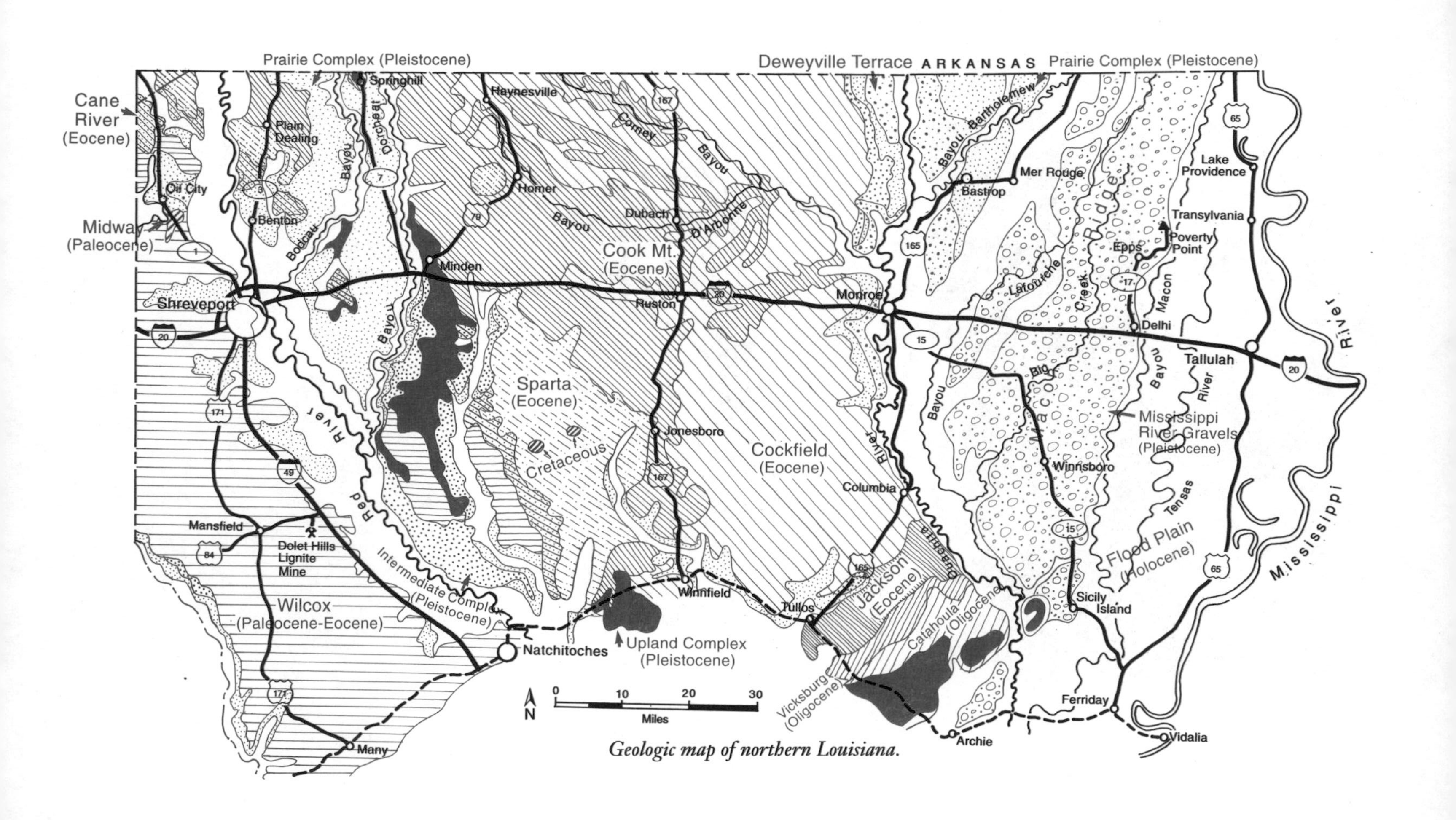

Geologic map of northern Louisiana.

Northern Louisiana

Oldest Rocks, Highest Hills

Enjoy the beauty and colorful hills in the western half of northern Louisiana, where you will find occasional excellent roadcuts and outcrops that reveal the rocks. Pay careful attention to subtle changes in elevation and topography, because any change could be telling you something about the rocks underfoot. The flat terrain of the Mississippi River valley, for example, stands in marked contrast to the abrupt edge of the Eocene uplands west of Monroe.

Louisiana has no mountains; the highest spot in the state is in the northwestern corner, east of Shreveport and just south of Interstate 20. The lack of real mountains has not deterred residents from giving their highest hill, 535 feet above sea level, the lofty title of Driskill Mountain.

Building the Youngest State

Mountains or not, the high, red hills of northern Louisiana stand in brilliant contrast to sea level marshes and delta plains that typify so much of the state. The oldest rocks are also here. North of Shreveport, rimming Caddo Lake, are 60-million-year-old shales that date back to Paleocene time, just after the demise of the dinosaurs. All other rocks and sediments in Louisiana are younger than these shales, making Louisiana one of the two youngest geologic states in the union, an honor it shares with Florida.

None of Louisiana stood above sea level at the end of Cretaceous time. Everything we now know as Louisiana is mud, sand, gravel, and silt that the ancestors of the Mississippi River added to North America during the past 60 million years.

In northern Louisiana, the oldest bands of sediment fan southward from Caddo Lake, overlaid by progressively younger sediment bands toward the south and east. It's easy to see the pattern of these bands on the geologic map of northern Louisiana, even though buried subsurface structures modify the pattern somewhat, especially around Shreveport.

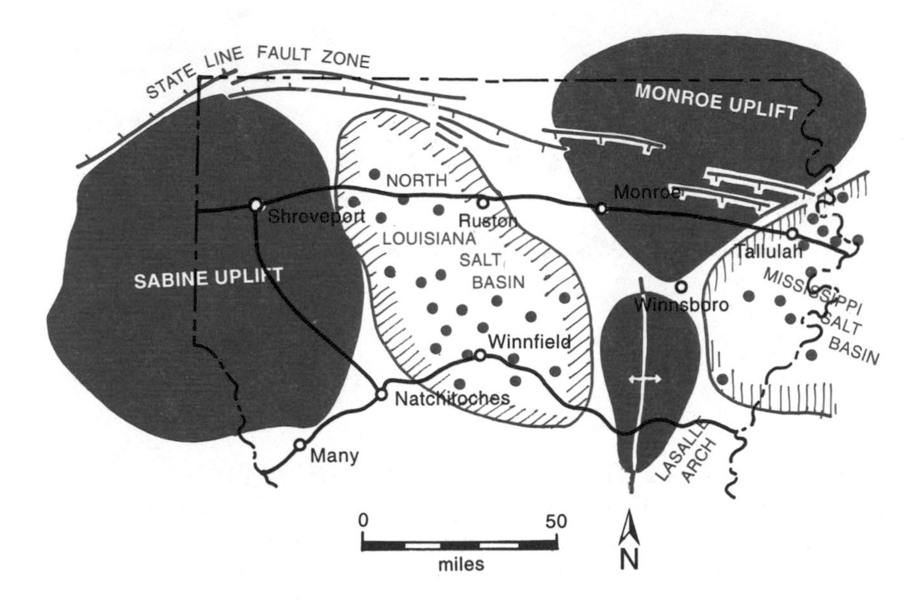

Map of the structures lying beneath the surface in northern Louisiana. Dots show salt domes in the North Louisiana salt dome basin and the Mississippi salt basin. The tops of most salt domes lie deeply buried, but a few have pushed close enough to the surface to create topographic domes. —Salvador, 1991

Uplifts and Basins

The Wilcox formation of the Paleocene and Eocene ages makes a curved pattern as it wraps around the margins of the Sabine uplift, a giant subsurface fold with a flat top that lies under the northwest corner of Louisiana and extends into East Texas. Most of Louisiana's subsurface structures are collapse features related to the expansion and sinking of the Gulf of Mexico. The Sabine uplift is an oddity along the Gulf Coast. It stands high, an arching fold probably pushed up by molten magma in Triassic time and later pushed up farther by the same westward forces of North American plate motion that shoved the Rocky Mountains skyward.

Its rise is one reason the uplift formed traps for oil and gas; early Louisiana production was found in these structures, especially around Caddo Lake. The giant East Texas oil field lies just across the border on the west flank of the Sabine uplift. Another big subsurface dome, the Monroe uplift, occupies the northeastern corner of the state. It hosts Louisiana's largest onshore gas field.

Between the two uplifts is a sag filled with salt pillars appropriately called the North Louisiana salt dome basin. A few of the salt pillars come quite close to the surface and form topographic domes. Two salt domes about 30 miles southeast of Shreveport have actually pushed Cretaceous marine limestone rocks to the surface. You could say these are the oldest naturally occurring rocks in Louisiana, but they are somewhat anomalous, and they are not in their original depositional position. Sneaking over the state line from Mississippi is a similar salt dome basin called the Mississippi salt basin. A portion of it lies beneath the Mississippi River floodplain south of Tallulah.

Except for the wraparound pattern of the Wilcox formation, the geologic map shows little surface evidence of the buried structures, either domes or sags. What is known about them comes from piecing together information from hundreds of oil and gas wells, a painstaking job that occupied geologists for many years.

Rivers

During the past few million years, the Red, Ouachita, and Mississippi Rivers have coursed across the northern Louisiana landscape, cutting deep valleys and filling them, all the while carving the older layers of sediment into hills and valleys. These rivers dominate the topography of northern Louisiana, and they played an equally important role in its settlement, transportation, and agriculture since prehistoric time.

Interstate 20
Texas—Shreveport—Monroe—Vicksburg
190 miles

This northern freeway follows a pleasant transect across Louisiana between Texas and Mississippi. It crosses many of the state's geologic elements, including the red uplands on the west, Macon Ridge, and the Mississippi floodplain on the east. The highway crosses all the major rivers: the Red, Ouachita, Tensas, and Mississippi. Beneath the surface are, from west to east, the Sabine uplift, North Louisiana salt basin, Monroe uplift, and a corner of the Mississippi salt basin. The variation in elevation here is the greatest in Louisiana, as the expressway passes near Driskill Mountain and dips to about 50 feet above sea level at the Mississippi River.

Between the Texas border and Shreveport, I-20 rolls across forest and farmland entirely underlaid by the pale sands and clays of the Paleocene and Eocene Wilcox formation, still lying nearly as flat as when they were deposited. Infrequent freeway cuts are grassed, hiding much of the Wilcox formation.

At Shreveport, the Red River cuts deeply into the Wilcox formation, and you can see the hills and elevation drop through town as the road descends toward the river. Shreveport is on the high west bank of the Red River. It owes its name to riverboat Captain Henry Shreve, who in 1833 became famous by heading a government crew to remove 165 miles of natural log jam from the Red River, opening it to the chugging paddle-wheel steamboats.

The Red River hugs the west side of its floodplain, which is eight miles wide and noted for its crimson soils, actually red muds eroded from red rock formations upstream. Look for the flood control levees that stand conspicuously above the otherwise featureless floodplain. They testify to frequent rises and falls in the river.

Eight miles east of Shreveport, the road abruptly climbs onto the surface of the Prairie Complex at the eastern edge of the modern Red River floodplain. The geologic map shows a patch of Wilcox and Eocene Cane River formations about five miles wide around Fillmore, but you hardly see them along I-20.

The road level between Fillmore and Dixie Inn is on an upper level of the Prairie Complex. These deposits are part of an abandoned segment of the Red River valley. The Pleistocene Red River shifted west about 15 miles to its present valley; Bayou Dorcheat now occupies the old valley. It is not clear why the Red River moved—whether as a result of stream erosion or perhaps because of movement on the Sabine uplift.

Deposits of the Upland Complex border the east bank of Bayou Dorcheat at Minden. Sandstone lenses of the Eocene Cook Mountain formation support the rolling and piney red hills east of Minden, where the road passes just north of a band of older Eocene Sparta formation. The geologic map shows bands of Eocene formations trending north and south in this area, with the oldest in the west and the youngest in the east. Farther

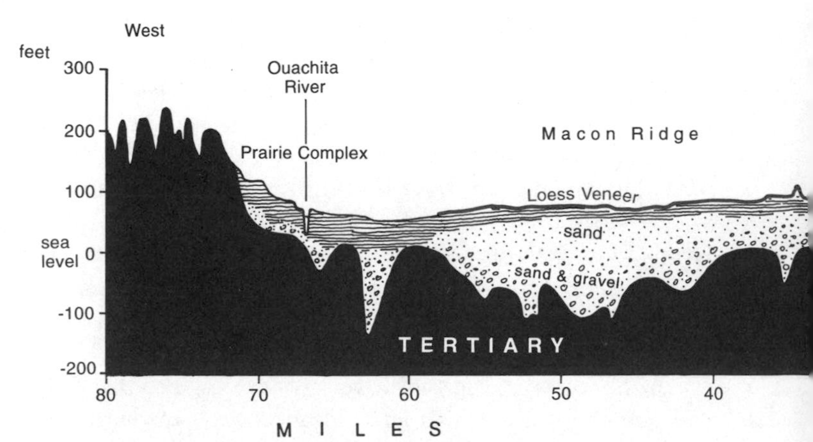

Profile along I-20, showing dimensions and nature of fill in the Mississippi-Ouachita River valley. —Saucier and Autin, 1990

south, the bands arc westward to outline the subsurface Sabine uplift. Watch between Minden and Arcadia for bankside exposures of Cook Mountain sandstones and the colorful orange and red soils.

Rolling hills continue between Arcadia and Ruston, but the soil is a bit lighter, and the rocks are the younger Eocene Cockfield formation. Watch for slides of soil on the steep slopes of several of the overpass banks.

The main rock between Ruston and Monroe is Cockfield formation, though the expressway dips into the valley of Bayou Choudrant between Ruston and Calhoun, where rocks of the lower Cook Mountain formation are at the surface.

The red hills give way to terrace levels west of Monroe, as the Tertiary rocks finally dip under, and are cut by, the Ouachita and Mississippi Rivers. About a mile of Prairie Complex is succeeded by another mile of lower Deweyville terrace on the west bank of the Ouachita River valley. The Ouachita River runs along the west side of its broad floodplain, where miles of levees break its flatness and protect its rich bottomland farms from most floods. The combined floodplain of the Ouachita River and Bayou Lafourche extends unbroken east of Monroe for 13 miles.

Look carefully for a subtle but recognizable change in the topography eastward to Delhi, where the highway crosses Macon Ridge. Macon Ridge is a pile of sand and gravel, glacial outwash laid down by the ancestral Mississippi River during Pleistocene time. Between Bayou Macon and Tallulah, the road crosses the Tensas River National Wildlife Refuge. This area is an abandoned Mississippi River meander belt and swamp.

Between Delhi and the Mississippi River, I-20 traverses the Mississippi River floodplain, 35 miles wide. It is flat and agriculturally rich. But the

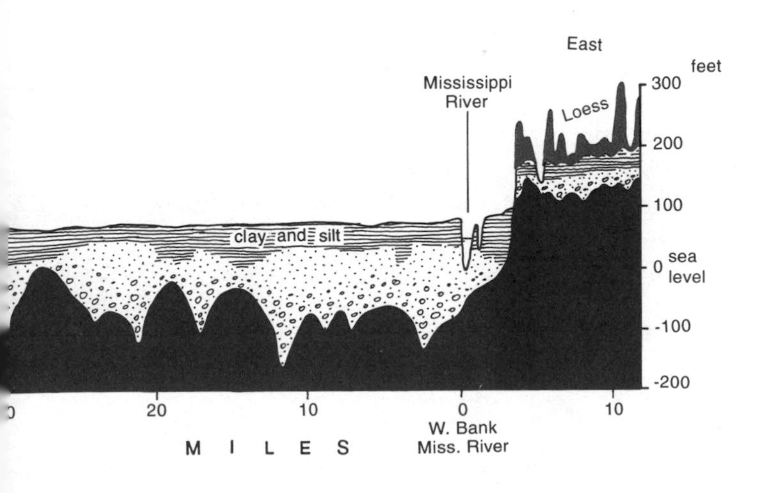

Mississippi River from high bluff at Vicksburg, Mississippi, looking across river bend to Louisiana.

very floods that used to bring new soil as they periodically inundated the land are now checked behind massive levees. When unusually high floodwaters bypass or breach levees, they deliver new soil to the floodplain while damaging farms and destroying houses.

A profile of the Mississippi Valley shows that from 100 to 200 feet of river sediment fills the modern valley, which was trenched into underlying Tertiary rocks during ice ages, when sea level was much lower than today. The river eroded to adjust its gradient to meet the sea at a more distant and lower level. As sea level rose when the ice melted at the end of an ice age, the Mississippi River dumped its sediment load higher in its course to again adjust the gradient to match a shorter distance to a higher sea level.

Watch as you drive over the high bridge to Vicksburg for the Mississippi River's broad stretch, the levees, the low topography on the inside bend of the river on the Louisiana side, and the high east bank, where the river relentlessly slashes laterally into the Tertiary rocks and loess deposits beneath Vicksburg.

Interstate 49
Shreveport—Natchitoches
66 miles

A View of the Wilcox Formation

The expressway follows the Red River, but stays mostly on the westerly high ground where the Paleocene and Eocene Wilcox group of formations are at the surface. Here and there the lanes dip into side drainages and onto the Red River floodplain. The difference between Wilcox and floodplain topography is obvious.

The numerous cuts and erosion patches along this new road are delightful little windows into the underlying rocks. How long they will last depends on how fast natural vegetation creeps in and whether the highway department regrades or reseeds these spots. But a few excellent steep cuts and even natural outcrops should provide good geologic vistas for many years.

For six miles south of Louisiana 3132 Loop, Prairie Complex deposits adjacent to Boggy and Cypress Bayous underlie I-49. Since the ancestral Red River had a hand in building these Pleistocene sediments, they are notably red, as are the modern floodplains. In the next 25 miles, the road bounces up and down across small tributaries to the Red River, all of which flow east. They dissect the Wilcox strata into a pleasantly hilly landscape. The various colors of the Wilcox rocks appear in many small cuts and erosion scars along the way.

Red iron concretions (Wilcox formation) formed around either roots or burrows. Along I-49, 24 miles south of Shreveport (Louisiana 3132 Loop). Quarter for scale.

Erosion rills in Wilcox formation claystones, along I-49 at Posey Road exit near Natchitoches.

The cuts on either side of the freeway are particularly good at the overpass four miles north of the US 84 crossover to Mansfield. An orange sand, full of rusty, nodular concretions, tops a brown claystone, which weathers gray and contains equally rusty concretion layers. The red concretions raining down the slope are more interesting than a casual look may reveal, because they are actually casts of Paleocene-age plant roots, and some may even be hardened marine animal burrows. Circulating groundwater deposited iron oxide in the casts, which makes them hard enough to resist erosion.

Freeway cuts about four miles south of the US 84 crossover nicely expose sand, silt, and clay in shades of white, pink, gray, and light orange, all part of the Wilcox formation. The road drops onto modern river alluvium for a few flat miles, then climbs into more hills eroded into the Wilcox formation. A big cut 15 miles south of US 84 exposes the Wilcox formation, where red soil tops gray clays and tan sands.

The variable character of the formation portrays discontinuous sedimentation and interwoven marine and nonmarine deposits. Concretion bands are well exposed along the southbound freeway lane north of the Ajax exit (Louisiana 174). Concretions are not part of the original deposit. They form much later, as groundwater flows through the deeply buried sand layers, depositing dissolved minerals in the microscopic spaces between sand grains.

The interstate highway cuts across a wide loop of Red River floodplain a few miles north of Louisiana 6, the Natchitoches exit. The loop is the old bed of Spanish Lake that was drained in the nineteenth century. The high earthen causeway the interstate follows was built to keep the road above floods. Look west to see the surrounding hills eroded in the Wilcox formation on the skyline above the red soils of the floodplain. Watch carefully for a natural sandstone outcrop perched on the southern bank edge of the floodplain, 5.5 miles north of Louisiana 6 on the west side of the southbound freeway lane. This is one of the best natural outcrops in Louisiana.

The brown sandstone contains large crossbeds, current ripples, and small clay pebbles, and rests on a gray claystone. The line between the sandstone and the claystone is very sharp. These features suggest that a river deposited this Wilcox sand in a channel on a clay floodplain. After deposition, the sediments were buried, and groundwater flowed through the sand, cementing it into solid rock. It also deposited the white calcite

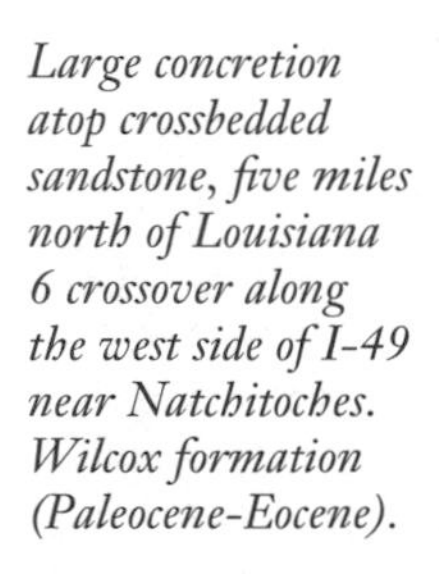

Large concretion atop crossbedded sandstone, five miles north of Louisiana 6 crossover along the west side of I-49 near Natchitoches. Wilcox formation (Paleocene-Eocene).

Black lignite bed above gray claystone, west side of I-49 at Posey Road exit near Natchitoches. Wilcox formation.

veins and cemented the marvelous round sandstone concretions that at first look like big boulders. This limy cementation made the sand hard, which is why it stands out here, resisting weathering and erosion.

West of the freeway, just south of the Posey Road exit 142, and one mile north of the Natchitoches exit/Louisiana 6, you can see a black bed of lignite exposed in a roadcut, another part of the Wilcox formation. It is about a foot thick and still preserves some plant fragment impressions. Look for eroded pillars and pedestals in the next roadcut to the south.

Dolet Hills Lignite Mine and Power Plant

In the Dolet Hills west of I-49, lignite from the Paleocene Wilcox formation is mined and burned to generate electricity. The Dolet Hills are the highest part of the Sabine uplift, and the mine is on the southwest dipping flank of the Sabine dome. To reach the mine and plant, take US 84 west from I-49 for 2 miles, then turn south on the Naborton Road and drive another 2.5 miles. Good roadcuts near the power plant display alternating beds of sandstone, siltstone, clay, and thin lignite stringers, all typical of the Wilcox formation in this area.

Lignite is very soft coal, brownish to black, just a step up from peat, two steps from compressed plant material. The Dolet Hills lignite is in the lower Wilcox formation among a series of sandstone, siltstone, and mudstone beds originally deposited in bays, lakes, streams, and freshwater swamps lying between distributary streams on a delta. Fossil plants in the lignite show that the precursor peat was deposited in an upper delta forest,

Small erosion pedestal in Wilcox formation claystone. Penny for scale at base of pedestal. One mile north of Louisiana 6 crossover along I-49.

Dragline at Dolet Hills lignite mine. —Williamson, 1987

in a freshwater swamp. Three lignite beds are mined, each from five to ten feet thick, with an estimated 140 million tons of reserves.

In the early 1950s, companies studied the feasibility of using lignite as boiler fuel to generate electricity, but the idea was abandoned because lignite could not then compete with cheap natural gas; by the 1970s, however, lignite became a cost-effective fuel. The Dolet Hills mine and power plant went into operation in 1985. A huge dragline crane and bucket mine about 2.5 million tons of lignite per year, monstrous trucks haul it 2 miles to a crusher, and a conveyor belt system carries it 7.5 miles to the generating plant. Reclamation and erosion control are part of the mining process; the mined area is continuously backfilled, graded, and revegetated, so no big pit ever exists. The power plant, piles of crushed lignite, and the roadcuts in the Wilcox are the features to see on this short side trip.

US 65
Tallulah—Ferriday
60 miles

This river route is entirely on natural levee and alluvial deposits of the Mississippi River. It is a straight road, skirting west of the river's bends and curves and oxbow lakes. The rich river deposits support vast tracts of cotton fields, punctuated by stands of swampland forest and palmetto groves. During the Civil War siege of Vicksburg in 1863, General Grant marched his troops along this route to avoid Confederate shelling as he established his winter quarters at a plantation southeast of Newellton, on Lake St. Joseph.

The road also crosses the west end of the Mississippi salt basin, though you see no evidence of this deep structural phenomenon at road level. The Mississippi salt basin was an outlying pocket that bordered the larger Gulf Coast salt basin in Jurassic time. Enough salt accumulated in this smaller basin to make a number of sizable salt domes when the burden of overlying sediment forced the lighter and more mobile salt to rise, like great globs of grease rising through water. The story of the subsurface rocks is gleaned from oil well drilling.

Look to the east for the artificial levees, the highest landmarks in the area. They are most conspicuous about 12 miles south of Tallulah and about 10 miles south of Newellton.

Four large oxbow lakes are along Highway 65: Lake St. Joseph and Lake Bruin lie between Newellton and St. Joseph, a historic nineteenth-century river town. Lake St. John is farther south, near Waterproof, and Lake Concordia is east of Ferriday. Early Indians were attracted to these lakes because they could live high and dry on the surrounding natural levees while fishing and hunting the lakes. Plantations were later located around the oxbows for similar reasons, in addition to the benefits of rich farmland and access to shipping on the nearby main river channel. Recreation now draws people to these pleasant oxbow lakes.

US 65
Tallulah—Lake Providence—Arkansas
45 miles

This byway along the Mississippi River creates an appreciation for the vastness of its floodplain and its natural levees. The road follows a route just east of enormous man-made levees, and the natural levees stretch away to the west. Their gray and tan clays and silts, the residue of ancient floods, make the rich soils that stretch to the western horizon. These deposits were laid down during the few thousand years since the last ice age.

Duckweed-covered abandoned channel in swampy bottomland along US 65, 38 miles south of I-20.

Ten miles south of the Arkansas border, the road loops around Lake Providence, an abandoned meander loop or oxbow lake, that was once the main channel of the Mississippi River. The town of Lake Providence is a pleasant resort community built on the natural levee. A careful look at this oxbow will give you a good idea of the size of the Mississippi's meanders.

When General Grant besieged Vicksburg, his engineers started a canal connecting Lake Providence with the river to allow Union gunboats to bypass shelling from Confederate shore batteries in Vicksburg. It was never completed. You can see a remnant of this old canal in the dip in topography next to the highway in town.

The port of Lake Providence, a major port on the Mississippi River, is a couple of miles south of Lake Providence. Six miles farther south, look for the railroad line climbing the side of the levee on its way to another river port. Watch for the bat on the water tower at Transylvania.

In Tallulah, Highway 65 follows Roundaway Bayou for a short distance north of I-20. It is another abandoned Mississippi River channel, now nearly filled with sediment and vegetation.

Brown claystone, weathered gray, over thin lignite bed of the Eocene Cook Mountain formation. A layer of hard iron concretions is visible at the base of the slope. Location is on the east side of Homer dome, three miles west of Homer.

79

US 79
Minden—Homer—Haynesville—Arkansas
35 miles

All the rocks along this trek were laid down during Eocene time. Ninety percent of the road length is on the Cook Mountain formation, while the northern 10 percent between Haynesville and the Arkansas border is on the Cockfield formation. The line between the two formations is gradational because both were deposited on a rapidly building delta. So the line between them is more a matter of interpretation than of hard fact—no two geologists would draw it in the same place on a map.

The rolling hills between Minden and Homer are the result of stream erosion of the Cook Mountain formation, which is about 50 million years old. Many small roadcuts in orange sandstone appear along the way. Quite good exposures exist six miles south of Homer, where sandstone weathered red overlies gray claystone. Abundant dark brown ironstone concretions and nodules weathered from the sandstone scatter across the slope. Yellow bands and stringers of limonite, an iron oxide mineral that contains water, lace through the mottled claystone.

Sedimentary rocks of the Cook Mountain formation poke from small roadcuts widely scattered between Homer and Haynesville. Between Haynesville and Arkansas, rocks of the Cockfield formation underlie the landscape, but you see very little of them. Haynesville is an oil town, centered in the Haynesville field. Watch for the pumps and the steel oil derrick in the Claiborne Parish fairground.

The circular patch on the geologic map west of Homer is a dome of Eocene Sparta formation that was pushed up above a salt dome. It is older than the other Eocene formations in the area.

A short side trip on Louisiana 2 west of Homer reveals a very good outcrop of the Cook Mountain formation three miles from town, near the base of a large hill. A thin bed of black lignite and thin layers full of yellowish ironstone concretions are mixed with the gray claystone. This outcrop is unusual because claystones are soft, weather easily, and are rarely exposed in the wet, nearly tropical climate of Louisiana. A fault at the base of the hill raises the Sparta formation above the level of the surrounding Cook Mountain formation. White and pink siltstone and sandstone beds of the Sparta formation are exposed in small erosion scars along the road at the top of the hill.

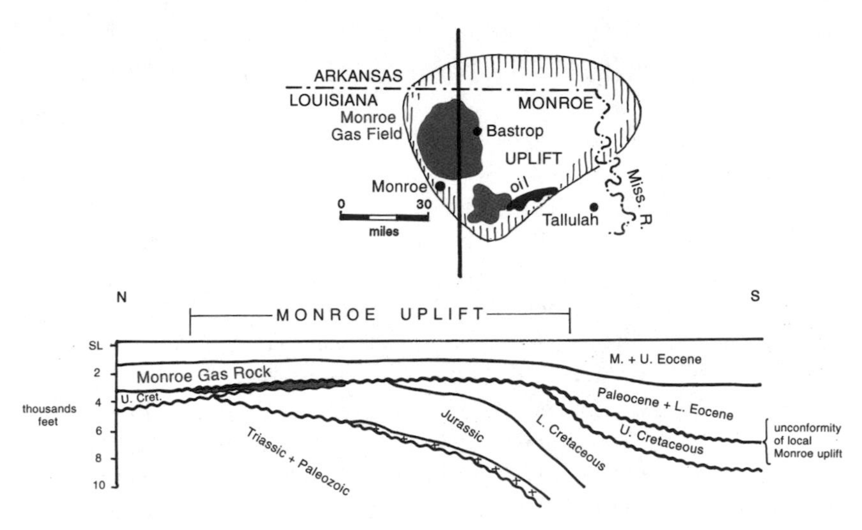

Map and cross section of Monroe uplift and Monroe gas field. —Ewing, 1991

US 165
Monroe—Bastrop—Arkansas
50 miles

Most of this road follows the modern floodplain level of the Ouachita River, Bayou Coulee, and Bayou Bonne Idee. The exception is a short jaunt across the pleasantly rolling Bastrop Hills, which separate the drainages of the Ouachita River and the two bayous.

The explorer Hernando de Soto crossed the Ouachita River at Monroe in 1542. The French built a trading post on the site in 1719, and the Spanish built Fort Miro in 1791. The Monroe uplift, named for the town, is a major subsurface structure in this northeastern corner of Louisiana. It is the location of the giant Monroe gas field, Louisiana's largest.

The Ouachita River is west of the highway where it follows its floodplain between Monroe and Bastrop. The smaller Bayou de Siard hugs the east side of the floodplain; it flows in a former channel of the Ouachita River. Watch four miles north of I-20 for the levees that protect the houses east of the road. A bit farther north, classic backswamp and waterway scenery surround the road. Though occasional tan soils of the Ouachita River appear in fields, the underlying sediment fill is mainly gravelly layers that the Mississippi River laid down as braided stream deposits, probably during ice ages when the river ran high, full of outwash debris from melting glaciers.

The highway climbs through hills into Bastrop. The Bastrop Hills separate the Ouachita River drainage on the west from Bayou Coulee and Bayou Bonne Idee to the east. Their southern end is part of the Deweyville terrace, the northern half of the older Prairie Complex. Highway 165 continues east of Bastrop for four miles on surprisingly hilly Prairie Complex. Loess on the Bastrop Hills accounts for some of the topography. Normally the Prairie surface is fairly flat with only minor stream dissection.

You will have no trouble spotting the edge of the Prairie Complex, where the road drops abruptly from a bluff to the modern floodplain level on the way to Mer Rouge. Between Mer Rouge and the Mississippi River, Highway 165 is on flat floodplain surface drained by Bayou Bonne Idee, Bayou Coulee, and Bayou Galion, old channels of the Mississippi River.

Look in Bonita for distinctive ridges and swales that tell of sand deposition on a large river point bar.

US 165
Monroe—Columbia—Tullos
57 miles

The road between Monroe and Tullos crosses two distinct landscapes: rich farmland dominates the scene between Monroe and Columbia as the highway follows the flat floodplain of the Ouachita River; the road between Columbia and Tullos weaves through hills eroded in Tertiary rocks dissected by entrenched streams.

The road between Monroe and Columbia travels along the natural levee surface adjacent to the Ouachita River. Light tan soils in fields are characteristic of the Ouachita River, in contrast to the red soils along the Red River and the gray soils along the Mississippi River.

Near Columbia, man-made levees rise 15 to 20 feet above the floodplain, nowhere near the stature of the bulwarks along the Mississippi River. The levee height reflects the amount of water the Ouachita River can carry in its floods, which do not begin to match the huge flows of the Mississippi.

Four miles north of Columbia, the highway curves around an oxbow lake, clearly visible next to the road. The scenery changes abruptly at Columbia, as Highway 165 ascends through nearly vertical bluffs eroded in Eocene Cockfield formation, too densely covered with vegetation to show many rocks. This rough topography illustrates how much the Ouachita River has eroded the older rocks. The bluff obviously defines the western edge of the Ouachita River floodplain.

Rocks of the Cockfield formation lie beneath the dissected terrain between Columbia and Burlington, but you can't see them. The road between Burlington and Tullos crosses deposits of the Prairie and Intermediate Complexes.

Tullos stands on the northern edge of a band of Eocene sedimentary rocks that belong to the Jackson group of formations. They are barely visible in exposures of tan sand along the road near town. Clays rich in lignite, bands of sand stained brown by iron oxide, and beds full of fossils are typical of the Jackson formation. Oil pumps at both ends of town produce from a small field.

US 167
Ruston—Dubach—Bernice—Arkansas
34 miles

This highway passes good roadcuts and outcrops of the brightly colored Eocene rocks that lie beneath much of northern Louisiana. On hilltops you will see brownish red weathered sandstones of the Cockfield formation, which rests on lighter brownish tan sandstones and siltstones of the Cook Mountain formation, exposed lower down in the valleys and stream cuts.

Red-weathered sandstone (Cockfield Mountain formation), 4.5 miles north of I-20 on US 167. Soft sand sediment bent around fossil log at left of center. Three-inch knife for scale (at arrow).

The landscape along the southern half of the road is hilly, dissected by small streams that wind eastward to join the Ouachita River near Monroe. Most of the rock exposures are Cook Mountain formation, though a few outcrops of the Cockfield formation are on hilltops between Ruston and Dubach. The terrain flattens north of Dubach as fewer streams cut through the depositional surface of the Cockfield formation. Most of the rocks north of Dubach belong to the Cockfield formation, except for narrow strips of Cook Mountain formation exposed in the stream valleys.

Watch for a nice red roadcut in Cockfield sandstone east of the road four miles north of Ruston. Thin beds of crossbedded sandstone contain white pebbles weathered to clay and deep red ironstone concretions. The red iron oxide is the rusted remnant of the green mineral glauconite, which is rich in iron. Glauconite typically forms on shallow sea floors, apparently from fecal pellets of small animals. Tree and shrub roots break the rocks at the top of the cut, contributing their part to weathering and erosion.

An outstanding roadcut exposes the Eocene Cockfield formation just a few yards east of US 167, at Rabb Road. Sandstone with crossbeds, big ironstone concretions, clay pebbles, and glauconite weathered red are all evident.

At 6.5 miles north of Ruston, US 167 dips into a small creek where yellowish sand of the Cook Mountain formation appears in a cut at the south end of the bridge. The small exposures of orange sandstone in yards and along the road in Dubach are in the Cook Mountain formation. On the southern edge of Bayou D'Arbonne, three miles north of Dubach, a roadcut exposes orange rocks in the Cook Mountain formation. Horizontal layers of fine sand are sandwiched between layers of gray clay. The clay weathers red on the surface, but if you dig a little you'll expose the gray.

Few rocks show up in the flat country north of Bernice, but a pit at the northern edge of town produces sand from the Cockfield formation. The orange and tan sand, as seen from the road, nicely represents the formation.

The countryside flattens out between Bernice and the Arkansas line, and the road rolls along on the depositional surface of the Eocene Cockfield formation.

US 167
Ruston—Jonesboro—Winnfield
48 miles

This is another Eocene run, mostly on Cook Mountain formation, though the road crosses about three miles of Cockfield formation just south of Ruston. Red soils and rolling hills are characteristic of the Cook Mountain formation along the highway between Ruston and Jonesboro, along an upper tributary of the Dugdemona River.

The Cook Mountain outcrop belt arcs from north to south. Highway 167 is at its center south of Jonesboro, where it is about two miles wide. Sandstone of the older Sparta formation lies to the west, the younger Cockfield formation is to the east. All three formations tilt gently down to the east on the flank of the Sabine uplift.

The road continues on a north-south path between Jonesboro and Winnfield, still in gently rolling hills eroded into the brownish tan Cook Mountain formation. The Prothro and Rayburns salt domes bring upper Cretaceous limestones and Paleocene rocks of the Midway formation to the surface in two spots: one is 14 miles west of Jonesboro on Louisiana 4, the other is on Louisiana 9 south of Lucky. You can get a much better view of salt dome caprocks at Winnfield, though no older bedded sedimentary rocks are exposed there.

The road crosses the Big Creek tributary of the Dugdemona River south of Wyett, still in rolling hills eroded on the Cook Mountain formation. Ten miles north of Winnfield, an old quarry on private land east of the road exposes reddish brown sandstone in the Cook Mountain formation. About five miles north of Winnfield, just south of Tannehill, the highway steps down onto a Prairie Complex level bordering the Dugdemona River. Then it enters the modern, swampy floodplain of the river, before climbing onto Prairie Complex for a mile or two. The highway proceeds into Winnfield, the birthplace of governors Huey Long, Earl Long, and O. K. Allen.

US 171
Shreveport—Mansfield—Many
80 miles

The entire route crosses rocks of the Paleocene and Eocene Wilcox group of formations. South of Shreveport, US 171 tracks down the south flank of the Sabine uplift, but the tilt of the layers of rock is hardly noticeable. The Wilcox formation lies so nearly flat that it covers a broad area on the geologic map.

The Wilcox formation was deposited in a delta setting about 55 to 60 million years ago. The Paleocene Midway group is older, but appears only in a small area around Caddo Lake. Clay beds and glauconite deposited in seawater are mixed with lignite, sand, and clay deposited above sea level. This variety gives the Wilcox formation its varied appearance and provides the evidence for its deposition on a delta. The lignite is thick enough to mine in the Dolet Hills, east of Mansfield. The mix of delta sands and clays rich in organic matter is ideal for forming oil and gas. The Wilcox formation produces large amounts of oil and gas farther south, where the formation is well below the surface.

Low, rolling farmland and bayou crossings typify the scenery. Boggy Bayou and Cypress Bayou north of Mansfield are tributaries of the Red River, whereas Bayous Patricio, San Miguel, and La Nana, between Mansfield and Many, feed the Sabine River to the west.

Geologists have mapped a number of faults that trend northeast to southwest between Many and Converse. They slice the Wilcox formation, but the offsets within it are impossible to see from a passing car window.

Louisiana 1
Shreveport—Oil City—Arkansas
36 miles

The oldest sedimentary rocks in Louisiana and, amazingly, the oldest offshore oil wells are tucked away in this northwestern corner. Louisiana 1 angles northwest from I-220, initially on red floodplain soils of the Red River. It quickly climbs onto a narrow segment of Prairie Complex, then traverses 12 miles of rolling hills covered with orange soils that support pine trees. They are eroded into the Eocene and Paleocene Wilcox formations.

Around the south rim of Caddo Lake, and in the vicinity of Mooringsport and Louisiana 169, is a thin band of gray to black shale. This is the Paleocene Midway group of formations, the oldest normally deposited sedimentary rocks exposed at the surface in Louisiana. Paleocene Midway and even older Cretaceous rocks appear in a few other isolated spots, but only where salt domes have punched them to the surface.

Where the road crosses Caddo Lake, look for old wooden oil platforms, still in operation. The world's first offshore, or at least overwater, oil well was drilled in Caddo Lake in 1911.

1

Oil City, a few miles north of Caddo Lake, is the home of the Caddo Pine Island Oil and Historical Society Museum, which features the colorful history of early oil discovery in this area. Outside the museum is a replica of the drill rig of the first discovery well.

Louisiana 1 crosses the Prairie Complex between Oil City and Vivian, which is on the Wilcox formation. The highway crosses a patch of brownish red Cane River sediments of Eocene age about two miles north of Vivian, then follows the drainage of the Black River to the Arkansas border. Rodessa is judiciously located above the river on a high ridge of Sparta sandstone, an Eocene deposit.

Discovery Well—replica of first drill rig in Caddo Lake area. At Caddo Pine Island Oil and Historical Society Museum, Oil City.

Louisiana 3
Shreveport—Benton—Plain Dealing—Arkansas
26 miles

Three Eocene formations stacked in order north of Shreveport tip very gently northward off the north flank of the Sabine uplift. Louisiana 3 crosses the elevated rolling terrain of these units, as well as the distinctively flatter and lower modern Red River floodplain near Bossier City.

Red soils in farm fields between Bossier City and Benton are characteristic of the scarlet sediment the Red River deposited on its floodplain. Most of that sediment was eroded from red mudstones that were deposited in western Oklahoma and West Texas during Permian time.

The levees a few miles north of the US 80 crossover now control the Red River. A mile south of Benton, the highway rises to the Prairie Complex level and continues on this elevated, low-rolling terrace to Alden Bridge. You encounter the lowest Eocene unit, the Cane River formation, just north of the junction with Louisiana 160. Watch for a roadcut where a yellowish orange sandstone contains dark brown ironstone concretions, six inches and larger in diameter.

From three miles north of Alden Bridge to Plain Dealing, the road traverses the next highest Eocene unit, the Sparta formation, but you hardly see it. In most places the Sparta formation is white sandstone and gray claystone with thin stringers of lignite.

North of Plain Dealing, the brownish tan roadcuts and rolling landscape are typical of the third and highest Eocene unit, the Cook Mountain formation. Near the Arkansas border, the road drops onto the flatter Intermediate Complex for about two miles.

Louisiana 7
Dixie Inn—Cotton Valley—Springhill—Arkansas
30 miles

Along most of its length, Louisiana 7 follows the high ground between Bayou Dorcheat to the east and Bayou Bodcau to the west. Both are tributaries of the Red River. The two bayous are in an abandoned section of the Red River valley. The Red River moved west about 15 miles to its present valley during Pleistocene time, but it not clear whether that happened because of ordinary erosion or as a result of movements of the earth's crust. Old channels and natural levees that have been mapped on the Prairie surface are remnants of several distinct meander belts.

The road traces the edge of the Prairie Complex for four miles north of Dixie Inn, then maintains the same level, on road fill, as it crosses about a mile of low cypress backswamp of Bayou Dorcheat. Louisiana 7 crosses the subtly undulating surface of the Prairie Complex between there and Springhill.

Cotton Valley is evidently named for the cash crop of yesteryear. The big refinery south of town and the Cotton Valley oil field a few miles southeast of town testify to the deeper cash crop of the twentieth century.

Two miles south of Cullen, 23 miles north of Dixie Inn, a large quarry west of the road provides sand for construction aggregate. Tan siltstone and sandstone overlie reddish weathered claystone in a classic exposure of deposits of the Prairie Complex. The quarry is on private land, but the view from the highway gives an overall impression of the Prairie sediments.

You'll cross Upland Complex levels between Springhill and the state line; the geologic map shows a narrow band of Eocene Cook Mountain formation around them. Neither is obvious from Louisiana 7.

Louisiana 15
Monroe—Winnsboro—Sicily Island—Clayton
78 miles

Macon Ridge is the main feature of this crossing between Monroe and Sicily Island.

The road follows the floodplain of the Ouachita River five miles southeast of Monroe, then eases over a little remnant of Deweyville terrace and drops into the Lafourche drainage, a tributary of the Ouachita River. The Bayou Lafourche channel is in an abandoned course of the Arkansas River. Two miles east of Bayou Lafourche is the edge of a shoulder of Macon Ridge about four miles wide that the Boeuf River separated from the main ridge. After crossing two miles of the Boeuf River valley, the road climbs onto Macon Ridge and follows it all the way between Alto and Sicily Island.

Macon Ridge is a wedge of sand, silt, and gravel laid down by the Mississippi River during the early part of the last great ice age. It now separates the Ouachita and Mississippi Rivers, though the Ouachita River finally slashes through the southern end of these deposits near Sicily Island. Many archeological sites are along the raised edge of Macon Ridge, where villages could be safe from flooding, yet near water and abundant food in the adjacent swamps and rivers.

Between Baskin and Mangham, Louisiana 15 crosses Big Creek, a local stream that drains Macon Ridge from its start 25 miles north of I-20 to its mouth in the Ouachita Valley east of Columbia.

Sicily Island is on the eastern edge of Macon Ridge above Mississippi floodwater level. Five miles west of town along Louisiana 87 are the Sicily Island Hills, a wildlife management area. The Upland Complex caps these lovely hills, and the geologic map shows a ring of white Oligocene Catahoula sandstone surrounding them. A Pleistocene deposit of windblown loess on the Upland Complex is named the Sicily Island loess.

The geologic map shows that Sicily Island is a peculiar round erosional remnant at the northeastern end of a belt of Tertiary rocks that extends southwest. It is at the southern end of Macon Ridge. Sicily Island is a high area between the Ouachita and Mississippi Rivers. It is a Louisiana landmark, both geologically and archeologically.

The road drops to modern floodplain level east of the town of Sicily Island, passes over Bayou Funny Louis, and remains on natural levee and alluvium sediments to Clayton.

Louisiana 17, Louisiana 134, Louisiana 577
Delhi—Poverty Point State Commemorative Area
17 miles

This short trek takes you 17 miles from I-20, but 37 centuries back in time, to the Poverty Point State Commemorative Area, where a sophisticated group of prehistoric Indians built an impressive array of ceremonial, effigy, and dwelling mounds. A central mound with the map outline of a bird rises 70 feet. It was laboriously assembled of earth carried in baskets. Adjacent rings of concentric, semicircular earthworks were probably foundations for dwellings.

The site is on a bluff overlooking Bayou Macon, which lies in an abandoned meander belt of the Mississippi River. The bayou drains the modern backswamp along the margin of the broad Mississippi floodplain. Here on the bluff, the village and its croplands were safe above raging floods yet near a steady water supply and a rich food source in the abundant animals and plants of the nearby swampland.

The bluff is the edge of Macon Ridge, 20 miles wide and 100 miles long on the west side of the Mississippi Valley. The ridge dates back to the early part of the last ice age, when the Mississippi River deposited a train of terraced outwash deposits of gray to brown sand, silt, clay, and gravel along the western edge of its floodplain. Macon Ridge's stance above the floodplain was crucial to the selection of Poverty Point as a village and ceremonial site some 3,700 years ago.

The lack of rocks in this area forced the Poverty Point villagers to trade far and wide for flint to make their stone tools and weapons. Some of the stone they used came from the Ohio and Tennessee river valleys. Soapstone for pots came from the Appalachian foothills of Georgia and Alabama. Many North American groups at that time cooked by dropping hot rocks into clay or stone pots, but the Poverty Point people had no rocks, so they molded clay into cooking balls. Thousands have been found. Poverty Point is a classic example of how indigenous people used local geologic conditions to their advantage.

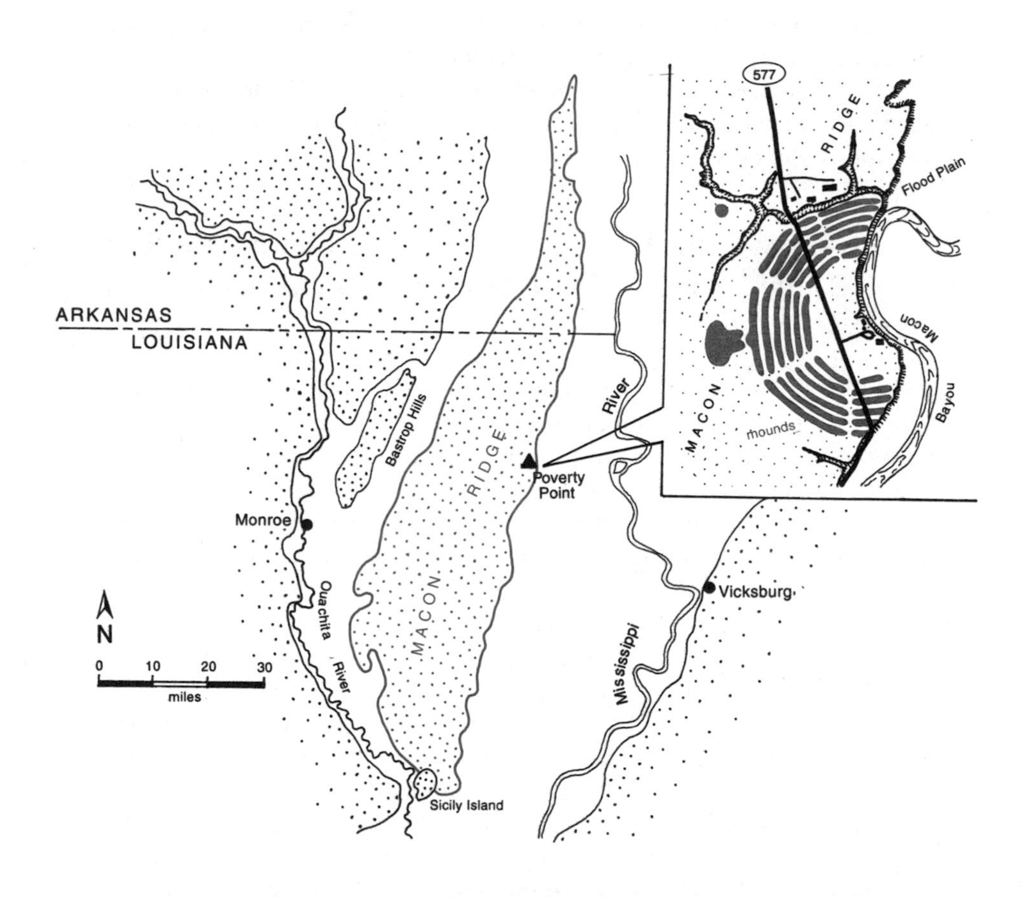

Large map shows location of Poverty Point village on edge of Macon Ridge, above flood level of Mississippi River valley. Inset map is detail of layout and mounds at Poverty Point. The builders used the local geology to their advantage!
—Louisiana Office of State Parks

An early snagboat. —Mills, 1978

Glossary

Aggregate: Material such as sand, gravel, or shells that is mixed with cement to make concrete.

Alluvium: Sediments laid down by recent rivers.

Anhydrite: The mineral form of calcium sulfate, but without the water that gypsum contains. Typically forms by evaporation of water.

Backswamp: Watery, low area away from river channels.

Barrier Island: Long, narrow coastal island, separated from the mainland by a shallow water lagoon or marsh.

Basin: Low area where sediments accumulate.

Bay: Open indentation of the coast, commonly where a rising sea level flooded the mouth of a river.

Bay Head Delta: Small delta built by a river emptying into the upper end of a bay, typically into a drowned river mouth.

Bayou: Louisiana term for a river, creek, or small stream, especially a sluggish one, winding through coastal swamps or deltas. Many are abandoned courses of larger streams.

Beach Ridge: Long mound of sand behind the beach; may occur as series of parallel ridges.

Beach Ridge Plain: Broad, flat area near the coast with a series of sandy beach ridges.

Bed: Smallest layer of sedimentary rock, usually a layer from one depositional event, such as a flood or hurricane.

Bedrock: Solid rock that underlies soil or loose surface material.

Braided Stream: Stream with an intricate network of interlaced channels; typically develops when sediment supply exceeds the stream's ability to move it all.

Calcareous: Rocks or soils that contain calcium carbonate, or calcite.

Calcite: The mineral form of calcium carbonate; main mineral in limestone.

Caprock: Impervious layer of anhydrite, gypsum, or limestone over a salt dome.

Chenier: An old beach ridge, generally covered with oak trees. It typically rises from coastal wetlands as a long ridge of sand that trends roughly parallel to the coast.

Chenier Plain: Broad, flat area occupied by cheniers and intervening muddy flats with marsh or swamp vegetation.

Claystone: Clay hardened into sedimentary rock.

Coastal Plain: Broad, low area bordering the coast; its strata are flat or slope gently seaward; usually a strip of recently emerged land.

Collapse Basin: More or less circular area in pile of sedimentary rocks that collapsed along curved faults as salt flowed out from under it.

Compaction: An early step in the process of changing sediment into rock. Occurs as load of the overlying sediments expels water and reduces pore space.

Concretion: A hard mass within a sedimentary rock. Commonly forms around a nucleus of a fossil, pebble, leaf, or shell.

Conglomerate: Sedimentary rock made up of gravel, pebbles, cobbles, or boulders.

Continental Slope: The part of the continental margin between the flat continental shelf and deep sea plain.

Crevasse: A break in a river levee.

Crevasse Splay: Fan of sediment dumped by river water as it spreads out and slows after spilling through a crevasse.

Crossbedding: Sedimentary layers deposited at an angle to the main horizontal beds.

Crust: The outer layer of the earth, many miles thick; includes both thick continental and thin oceanic crust.

Deformation: The process of folding, faulting, compressing, or extending rocks.

Delta: A deposit of sediment formed at the mouth of a river, either in a lake or ocean.

Delta Plain: Flat area formed along the coast by coalescing delta deposits.

Depositional Environment: The place where sediment is deposited, such as a river, delta, lake, or shallow sea.

Discharge: The volume of river water flowing past a point for a given time, such as cubic feet per second.

Distributary: Branches of a river that flow away from the main channel, as on a delta.

Drainage Basin: Area where all the streams gather water into a common trunk stream or river, for example, the total area of water collection of the Mississippi River.

Erosion: The wearing away of the landscape by natural forces of water, waves, ice, wind, or tides.

Erosion Rill: Small, nearly parallel channels, cut into steep slopes by flowing water.

Escarpment: An abrupt change in ground elevation.

Estuary: Tidal mouth of a river, where seawater and fresh water mix; forms when rising sea level drowns a river mouth.

Evaporite Mineral: Mineral precipitated as a result of evaporation, such as salt, anhydrite, and gypsum.

Fault: Fracture in rock along which one side moves up, down, or sideways past the other.

Floodplain: Flat area adjacent to a river channel that may be submerged when the river overflows its channel banks during a flood.

Floodway: An area confined by levees used to direct water away from a flooding river.

Formation: A distinctive and mappable body of rock, generally named for a geographic locality and usually for the main rock type, for example Catahoula sand; the fundamental sedimentary rock unit.

Fossil: Any preserved evidence of past life.

Glacier: Mass of moving ice, formed by compaction and recrystallization of snow.

Glauconite: A green mineral, a variety of mica that contains iron, formed into small pellets in shallow seawater where sedimentation is very slow.

Groundwater: Subsurface water contained in porous soil, sediment, or rock.

Gypsum: Soft, white mineral, watery calcium sulfate; common evaporative mineral along with anydrite and salt.

Headwaters: The source of a stream or river.

Hematite: Reddish iron oxide mineral; chief ore of iron.

Ice Age: A glacial episode during Pleistocene time.

Interbedded: Beds lying between or alternating with beds of a different kind, such as sand interbedded with clay.

Ironstone: Rock rich in iron oxide.

Levee: Artificial or natural bank confining a stream to its channel.

Lignite: Brownish to black, highly compressed plant material, between peat and bituminous coal in coal rank.

Limestone: Sedimentary rock primarily composed of the mineral calcite, calcium carbonate.

Limonite: Yellowish brown hydrous iron oxide mineral, rust.

Loess: A deposit of windblown silt.

Marsh: Treeless and periodically flooded ground.

Meander: Loops or bends in a river.

Metamorphic: Rock formed from older rock by recrystallization at high temperature and pressure.

Mudstone: Rock composed chiefly of mud.

Outcrop: Bedrock exposed at the earth's surface.

Overburden: Whatever rock lies over the rock you are interested in.

Oxbow Lake: An abandoned meander loop filled with water. The name derives from a fancied resemblance in map view to an ox yoke.

Peat: Partly decomposed plant remains; an early stage of coal development.

Petrified Wood: Fossil wood hardened by permeation with silica.

Petroleum: Crude oil or natural gas.

Point Bar: Area on the inside bend of a river meander.

Protection Levee: Artificial levee designed to protect land from flooding.

Pyrite: Mineral form of iron sulfide; fool's gold.

Quartz: Mineral form of silicon dioxide; most sand grains are quartz.

Radioactive Decay: Some elements, such as uranium, change into other elements; the rate of this change is constant. By measuring the amount of parent and daughter elements in a rock and knowing the decay rate, geologists can calculate the age of the rock.

Reef: A mound or ridge built in shallow marine water by organisms such as corals.

Reservoir (oil or gas): Subsurface rock with sufficient connected holes to store commercial volumes of oil or gas.

Remote Sensing: Finding out about the earth by using various instruments such as radar, gravity meter, or seismic recorder.

Ripple, Ripple Mark: Series of small ridges of sand produced when wind or water moves sediment.

River Mouth Bar: Broad pile of sediment deposited in front of a river mouth, in either ocean or lake.

Salt Basin: A region that contains large numbers of salt domes.

Salt Dome: Extremely long pillar of salt rock that rose from great depth to penetrate the overlying sediment; many of those along the Gulf Coast created traps for oil and gas in the tilted layers of sedimentary rocks above the dome and on its flanks.

Sand Dune: Pile of sand heaped up by the wind.

Sandstone: Sedimentary rock composed mainly of sand grains.

Sedimentary Rock: Sediment consolidated into solid rock.

Shale: Thinly layered rock composed of clay.

Shelf: Flat area on continent margin; can be dry land or covered by shallow ocean water, depending on sea level.

Silica: Silicon dioxide; makes up the mineral quartz.

Silt: Sediment composed of grains smaller than sand but bigger than clay.

Siltstone: Sedimentary rock composed of silt.

Soda Ash: Naturally occurring sodium carbonate.

Spillway: Control structure that lets water out of a flooded river into an area where the flood damage will be slight.

Spit: Narrow point of land jutting out into a body of water from the shore.

Strata: Layers of sedimentary rock.

Stratigraphy: The study of sedimentary rocks.

Subsidence: The sinking of a part of the earth's crust.

Swale: Low, swampy area, as between two river point-bar ridges or two chenier ridges.

Swamp: Low area shallowly flooded with fresh water that commonly supports trees, such as cypress and tupelo.

Terrace: A flat bench above present river level, a remnant of an old floodplain now partially eroded.

Tidal Flat: Flat, barren, or marshy area that is alternately flooded and drained as the tide rises and falls.

Tributary River or Stream: Any stream or river that flows into a larger river or stream.

Unconformity: A significant break or gap in the geologic record, caused by erosion or lack of deposition.

Volcanic Ash: Very fine shreds of lava that blow from a volcano and drift on the wind.

Credits

All photographs were taken by the author, and all maps and illustrations were drawn by the author specifically for this book.

Many illustrations are the author's own, others are modified and adapted from other geologists' published books and papers, which are cited in "References," and used with kind permission of authors and publishers. Credited illustrations are identified by the author's or source's name and publication date, in the caption of each illustration

Geologic maps are modified and adapted mainly from the Geologic Map of Louisiana 1:500,000, compiled by John. I. Snead and Richard P. McCulloh, published in 1984 by the Louisiana Geological Survey; and the Map of Quaternary Geology of the Lower Mississippi Valley 1:1,100,000, compiled by Roger T. Saucier and John I. Snead, published in 1989 by the Geological Society of America.

Source material for the text has been gleaned from the abundant geologic literature on Louisiana. Much information comes from guidebooks, pamphlets, and papers published by the Louisiana Geological Survey and other sources, which are listed in "Maps and Information" and "References." The format of this book is not conducive to citing every one of these literature sources, though most are listed in the "References;" the author gratefully acknowledges and recognizes the many geoscientists who have devoted careers to unraveling the geology of Louisiana and who have published their observations and ideas.

References

Autin, W. J. 1984. Geologic Significance of Land Subsidence at Jefferson Island, LA. *Trans. Gulf Coast Assoc. Geol. Socs.*, 34:293–309.

Autin, W. J. et al. 1986. The Florida Parishes of Southeast Louisiana. *Geol. Soc. Amer. Centennial Field Guide—SE Sect.*, pp. 419–23.

Autin, W. J., and C. J. John. 1990. Development of an Industrial Minerals Program for Louisiana's Chemical and Construction Industries. In Quaternary Non-glacial Geology. *Geol. Surv. Alabama, Circ. 161*, Industrial Mins. of SE United States, p. 92.

Autin, W. J., et al. 1991. Quaternary Geology of the Lower Mississippi Valley. In Quaternary Non-glacial Geology. *Geology of North America*, v. K–2. Geol. Soc. Amer., pp. 547–82.

Basement Map of North America. 1967. Amer. Assoc. Petroleum Geol. and U.S. Geol. Survey.

Belchic, H. C. 1960. The Winnfield Salt Dome, Winn Parish, LA. In *Shreveport Geol. Soc., Spring Field Trip Guidebook*, pp. 29–41.

Bennison, A. P. 1975. *Geologic Highway Map of the Southeastern Region*, Map no. 9. amer. Assoc. Petroleum Geol.

Boyd, R., J. Suter, and S. Penland. 1989. Sequence Stratigraphy of the Mississippi Delta. *Trans. Gulf Coast Assoc. Geol. Socs.*, 39:331.

Burgess, W. J. 1976. Geologic Evolution of the Mid-Continent and Gulf Coast Areas—A Plate-Tectonics View. *Trans. Gulf Coast Assoc. Geol. Socs.*, 26:132–43.

Burke, K. 1980. Intracontinental Rifts and Aulacogens. In *Continental Tectonics—Studies in Geophysics*. Washington, D.C.: Nat. Acad. Science. p. 44.

Coleman, J. M. 1976. *Deltas: Processes of Deposition and Models for Exploration*. Champaign, IL: Cont. Educ. Publ. Co. fig. 12.

Coleman, J. M.. 1988. Dynamic Changes and Processes in the Mississippi River Delta. *Geol. Soc. Amer. Bull.* 100:999–1015.

Davis, D. W. 1990. Living on the Edge: Louisiana's Marsh, Estuary and Barrier Island Population. *Trans. Gulf Coast Assoc. Geol. Socs.* 40:147–59.

Ewing, T. E. 1991. Structural Framework of the Gulf of Mexico Basin, Ch. 3. In The Gulf of Mexico Basin. *Geology of North America*, v. J, edited by A. Salvador. Geol. Soc. Amer.

Flowers, G. C. and W. C. Ishpording. 1990. Environmental Sedimentology of the Pontchartrain Estuary. *Trans. Gulf Coast Assoc. Geol. Socs.* 40:237–50

Gagliano, S. M., K. J. Meyer-Arendt, and K. M. Wicker. 1981. Land Loss in the Mississippi River Deltaic Plain. *Trans. Gulf Coast Assoc. Geol. Socs.* 31:295–300.

Gould, H. R., and E. McFarlan Jr. 1959. Recent Sediments of the North-Central Gulf Coast Plain. *Amer. Assoc. Petroleum Geol., Ann. Mtg. Guidebook.*

Halbouty, M. T. 1979. *Salt Domes, Gulf Region of the United States and Mexico*, 2nd ed. Houston: Gulf Publ. Co.

Hamblin, W. K. 1985. *The Earth's Dynamic Systems*, 4th ed. Minneapolis: Burgess Publ. Co., figs. 14.18, 14.22, 14.23.

Hoyt, J. H. 1969. Chenier versus Barrier, Genetic and Stratigraphic Distribution. *Amer. Assoc. Petroleum Geol. Bull.* 53:299–306.

Kupfer, D. H., and J. P. Morgan. 1976. Louisiana Delta Plain and Salt Domes. *New Orleans Geol. Soc. Field Trip Guidebook.*

Louisiana Geological Survey. 1980. *Map of Louisiana Salt Domes.*

Louisiana Geological Survey. 1981. *Oil and Gas Map of Louisiana.*

Louisiana Land and Exploration Co. 1990. *Louisiana's National Treasure.* pamphlet.

Louisiana Office of State Parks. *Trail Guide: Poverty Point State Commemorative Area.*

Manning, E. 1990. The Late Early Miocene Sabine River. *Trans. Gulf Coast Assoc. Geol. Socs.*, v. 40, pp. 531–549.

McGehee, E. L. 1983 Oil and Gas Fields and Salt Domes, Louisiana. *Resources Info. Series No. 1.* Louisiana Geol. Surv.

Mills, G. B. 1978. *Of Men and Rivers.* U.S. Army Corps Engrs. Vicksburg.

Otvos, E. G., Jr. 1978. New Orleans—South Hancock Holocene Barrier Trends and Origins of Lake Pontchartrain. *Trans. Gulf Coast Assoc. Geol. Socs.* 28:337–55.

Otvos, E. G., Jr. 1971. Relict Eolian Dunes and the Age of the "Prairie" Coastwise Terrace, Southeastern Louisiana. *Geol. Soc. Amer. Bull.* 82:175–358.

Pearson, C. E., and D. G. Hunter. 1993. Moncla Gap and the Red River Diversion. In *Quaternary Geology and Geoarcheology of the Lower Red River Valley, Southeast Cell Annual Field Trip Guidebook*, edited by W. J. Autin and C. E. Pearson. Friends of the Pleistocene.

Penland, S., and R. Boyd, eds. 1985. Transgressive Depositional Systems of the Mississippi River Delta Plain. *Louisiana Geol. Surv. Guidebook*, ser. no. 3, p. 83.

Penland, S. et al. 1986. The Bayou Lafourche Delta, Mississippi River Delta Plain, Louisiana. In *Centennial Field Guide*, v. 6, Southeast Section, edited by T. L. Neathery. Geol. Soc. Amer.

Roberts, H. H., R. D. Adams, and R. H. W. Cunningham. 1980. Evaluation of Sand-Dominant Subaerial Phase, Atchafalaya Delta, Louisiana. *Amer. Assoc. Petroleum Geol. Bull.* 64(2):264–79.

Roberts, W. H., III. 1982. Gulf Coast Magic. *Trans. Gulf Coast Assoc. Geol. Socs.* 32:205–15.

Salvador, A., ed. 1991. The Gulf of Mexico Basin. *Geology of North America*, v. J, Plate 2. Geol. Soc. Amer.

Sassen, R. 1990. Lower Tertiary and Upper Cretaceous Source Rocks in Louisiana and Mississippi. *Amer. Assoc. Petroleum Geol. Bull.* 74(6):857–78.

Saucier, R. T. 1986. Holocene Fluvial Landforms and Depositional Environments of the Lower Mississippi Valley. *Geol. Soc. Amer. Centennial Field Guide—SE Sect.* 409–12.

Saucier, R. T., and W. J. Autin. 1990. Lower Mississippi Valley Cross Sections. *Geology of North America*, v. K-2, plate 7. Geol. Soc. Amer.

Saucier, R. T., and J. I. Snead. 1989. Quaternary Geology of the Lower Mississippi Valley. *Geology of North America*, v. K–2, map. Geol. Soc. Amer.

Schumm, S. A., and G. R. Brackenridge. 1987. River Responses. In North America and Adjacent Oceans During the Last Deglaciation. *Geology of North America*, v. K-3, p. 229. Geol. Soc. Amer.

Scott, A. J. 1969. In Fisher, W. L., et al. *Delta Systems in the Exploration for Oil and Gas.* Bur. Econ. Geol., Univ. Texas, Austin.

Scruton, P. C. 1960. Delta Building and the Deltaic Sequence. In *Recent Sediments, Northwest Gulf of Mexico*, edited by F. P. Shepard et al. Amer. Assoc. Petroleum Geol.

Seglund, J. A. 1974. Collapse Fault Systems of Louisiana Gulf Coast. *Amer. Assoc. Petroleum Geol. Bull.* 58(12):2389–97.

Snowden, J. O., W. C. Ward, J. R. J. Studlick, and L. E. Rieg. 1980. Geology of Greater New Orleans. *New Orleans Geol. Soc. Guidebook.*

Spearing, D. R. 1991. *Roadside Geology of Texas.* Missoula, Mont.: Mountain Press.

Steward, H. L., and W. L. Berry. 1990a. Louisiana's Vanishing Coastal Wetlands. *Geotimes.* 35(6):19–21.

———. 1990b. *Louisiana's National Treasure.* pamphlet. New Orleans: La. Land and Explor. Co.

Stuart, C. J., and C. A. Caughey. 1976. Form and Composition of the Mississippi Fan. *Trans. Gulf Coast Assoc. Geol. Socs.* 26:333–41.

Suter, J. R., S. Penland, S. J. Williams, and J. L. Kindinger. 1988. Transgressive Evolution of the Chandeleur Islands, LA. *Trans. Gulf Coast Assoc. Geol. Socs.* 38:333.

U.S. Army Corps of Engineers. 1978. *Atchafalaya Basin.* pamphlet.

vanHeerden, I. L., and H. H. Roberts. 1980. The Atchafalaya Delta—Louisiana's New Prograding Coast. *Trans. Gulf Coast Assoc. Geol. Socs.* 30:497–506.

Walker, H. J., and J. M. Coleman. 1987. Atlantic and Gulf Coastal Provinces. In Geomorphic Systems of North America. *Geol. Soc. Amer. Centennial Spec.* 2:65, edited by W. L. Graf.

Williamson, D. R. 1987. *Lignites of Northwest Louisiana and the Dolet Hills Lignite Mine.* Paper presented at the 14th Ann. Lignite Symposium for the U.S. Dept. Energy, and Environmental and Coal Assocs., Reston, VA.

Wood, M. L., and J. L. Walper. 1974. The Evolution of the Interior Mesozoic Basin and the Gulf of Mexico. *Trans. Gulf Coast Assoc. Geol. Socs.* 24:31–41.

For readers interested in pursuing more of the intriguing story of the Atchafalaya Basin, the following are recommended:

Write for a free copy of "Atchafalaya Basin," a well-designed brochure available from the U.S. Army Corps of Engineers, P.O. Box 60267, New Orleans, LA 70160.

The delightfully lyrical book *The Control of Nature* by John McPhee (1989, New York: Noonday Press).

For readers interested in reading more about the fascinating geology and geologic history of New Orleans, this exceptionally readable booklet is recommended:

Geology of Greater New Orleans by J. O. Snowden, W. C. Ward, and J. R. J. Studlick (1980. New Orleans Geological Society, P.O. Box 52171, New Orleans, LA 70152). The booklet includes "Geologic Walking Tour of Downtown New Orleans" by L. E. Rieg.

Maps and Information

Geologic maps of Louisiana and detailed local and regional guidebooks are available from several sources. Write to the following organizations for a list of available publications and prices.

Louisiana Geological Survey (LGS)
University Station, Box G
Baton Rouge, LA 70803

Geological Society of America (GSA)
3300 Penrose Place, P.O. Box 9140
Boulder, CO 80301

American Association of Petroleum Geologists (AAPG)
AAPG Bookstore, P.O. Box 979
Tulsa, OK 74101

United States Geological Survey (USGS)
Branch of Distribution, Box 25286
Federal Center
Denver, CO 80225

Baton Rouge Geological Society
P.O. Box 19151, University Station
Baton Rouge, LA 70893

Lafayette Geological Society
P.O. Box 51896
Lafayette, LA 70505

New Orleans Geological Society
P.O. Box 52172,
New Orleans, LA 70152

Shreveport Geological Society
P.O. Box 750
Shreveport, LA 71162

Earth Enterprises
P.O. Box 672
Austin, TX 78767

A few key maps are:

Geologic Map of Louisiana
1:500,000—1984 (LGS)

Quaternary Geology of the Lower Mississippi Valley
1:1,100,000—1989 (GSA, LGS)

Geologic Highway Map,
Southeastern Region (includes Louisiana)
1" = 30 mi.—1975 (AAPG)

Parish Geologic Maps
each 1:62,500 (LGS)

Oil and Gas Map of Louisiana and
Offshore Louisiana Oil and Gas Map
1:380,160—1981 (LGS)

Louisiana Salt Domes
1" = 40 mi.—1980 (LGS)

Historical Shoreline Change
in the Northern Gulf of Mexico
1:2,000,000—1991 (LGS)

Recent Geologic Floodplain Deposits of Louisiana
1:500,000—1982 (LGS)

Index

About the Author

Darwin R. Spearing, author of *Roadside Geology of Texas* and coauthor of *Roadside Geology of Wyoming*, whetted his interest in the geology of Louisiana as a research geologist and exploration manager in the petroleum industry. Since then, he has been a museum designer and a national park ranger. Spearing writes from his home in Grand Lake, Colorado.